FOOD SCIENCE AND TECHNOLOGY

AVOCADO

CONSUMPTION AND HEALTH

FOOD SCIENCE AND TECHNOLOGY

Additional books and e-books in this series can be found on Nova's website under the Series tab.

FOOD SCIENCE AND TECHNOLOGY

AVOCADO

CONSUMPTION AND HEALTH

MARÍA GUIOMAR MELGAR LALANNE
AND
ALAN JAVIER HERNÁNDEZ ÁLVAREZ
EDITORS

nova
science publishers
New York

Additional color graphics may be available in the e-book version of this book.

Library of Congress Cataloging-in-Publication Data

Names: Guiomar Melgar Lalanne, María, editor. | Javier Hernández Álvarez, Alan, editor.
Title: Avocado consumption and health / María Guiomar Melgar Lalanne, editor, Basic Sciences Institute, Veracruzana University, Veracruz, Mexico, Alan Javier Hernandez Alvarez, editor, School of Food Science and Nutrition, University of Leeds, Leeds, UK.
Identifiers: LCCN 2020029129 (print) | LCCN 2020029130 (ebook) | ISBN 9781536182040 (hardcover) | ISBN 9781536182804 (adobe pdf)
Subjects: LCSH: Avocado in human nutrition. | Avocado--Therapeutic use.
Classification: LCC QP144.A77 A96 2020 (print) | LCC QP144.A77 (ebook) | DDC 634/.653--dc23
LC record available at https://lccn.loc.gov/2020029129
LC ebook record available at https://lccn.loc.gov/2020029130

Published by Nova Science Publishers, Inc. † New York

CONTENTS

INTRODUCTION

Lidia Dorantes-Álvarez*, PhD
National School of Biological Sciences (ENCB), National Polytechnic Institute (IPN), Mexico

Avocado fruit (*Persea americana* Mill) consumption has increased exponentially around the world, mainly due to its exquisite flavor and texture, but also to the accumulated scientific evidence showing that consumption of avocado reduces cholesterol and low-density lipoproteins while improving the fatty acid profile with its high content of monounsaturated oleic acid. Recently, more evidence of the beneficial effects of avocado products to overcome chronic diseases such as cardiovascular disorders has been reported.

The outstanding properties of avocado pulp rely on its unique composition of lipophilic and hydrophilic phytochemicals. Microscopic and chemical studies have shown that avocado idioblasts - cells where oil predominates- contain chlorophylls and carotenoids (α, β, neoxanthin, zeaxanthin, violaxanthin), phytosterols such as β -sitosterol, tocopherols and, squalene. Phenolic compounds are present in the aqueous and non-aqueous phases of avocado and are interesting fractions due to their

* Corresponding Author's Email: lidiadorantesa@gmail.com, ldoran@ipn.mx.

bioactive properties, such as antioxidant and antimicrobial. Particularly, anthocyanins and flavonoids have shown to contribute to health maintenance in cases of malnutrition. The unsaponifiable fractions of avocado have been used in the therapeutic management of arthrosis.

The book is mainly directed to academics, undergraduate and postgraduate students (including research students) in food science and technology, biotechnology, and bioengineering, who will find a selection of topics ranging from the nutritional characterization of raw avocado, bioactive compounds, and biotechnology to industrial applications. Professionals working in food science and technology research centers also may find this book useful.

This book presents a non-exhaustive, updated information of clinical and preclinical research that correlates the consumption of avocado and human health in Chapters 1 to 3: "Therapeutic and nutraceutical Potential of *Persea americana* fruits and their by-products"; "Avocado (*Persea americana* Mill.) potential as a source of bioactive compounds for human health and wellbeing"; and "Nutraceutical profile of avocado and health."

Moreover, Chapter 6 is devoted to reviewing relevant intellectual property through the review of "Patents on avocado products related to health" with application in pharmaceutical, nutraceutical, dietary supplements, and cosmetic industries. They are providing comprehensive data on the commercial avocado products available in the market, opening new alternatives for the development and enhancement of functional and nutraceutical ingredients from avocado.

Another aspect of the contents of this book is focused on the properties of phytochemicals extracts present in the avocado pulp, oil, and by-products peel and seeds, for example Chapter 9 "Antimicrobial activity of phenolic compounds present in virgin avocado oil", and Chapter 10 "Fatty acid profile of oil from Hass avocado seed (*Persea americana* Mill.) extracted by three different methods".

Some experimental data on different aspects of stability and quality of avocado products are presented in the Chapters 8 and 4: "Studies of application of electric fields technology to prevent the enzymatic browning of avocado catalyzed by polyphenol oxidase" and "Factors concerning the oxidative stability and shelf life of avocado oil."

The preservation of the three avocado horticultural varieties is also of concern, to create an *in vitro* germplasm bank using biotechnology and *in vitro* culture techniques, which may provide pathogen-free plants, this is presented in Chapter 5: "Biotechnology and *in vitro* culture for the sustainability of avocado agroecosystems (*Persea americana* Mill)."

The formulation of products is another aspect of this book. Guacamole is an avocado product widely consumed in some countries. It is basically avocado pureed or mashed with lemon, salt, and condiments, as shown in Chapter 7: "Characterization of guacamole formulated with two varieties of Yucatecan avocado", which analyses the sensory and microbial aspects of two formulations. Also, Chapter 11: "Chipotle chili oleoresin extracted in avocado oil as an additive in amaranth bars" presents a formulation using avocado oil to extract oleoresin of chili as an additive.

Furthermore, future research must be aimed at supplementary preclinical and clinical knowledge of bioactive compounds of avocado pulp, seeds, peels and leaves and its contribution to human health is expected; as well as new technologies to improve the yields of extraction of avocado phytochemicals, maintaining the properties of these valuable components.

Thanks to Editors Alan Javier Hernández-Alvarez PhD and Guiomar Melgar-Lalanne PhD for the invitation to write this presentation and to Nova Science Publishers for making the edition of this book possible.

ABOUT THE EDITORS

Guiomar Melgar-Lalanne is currently a researcher in the Basic Sciences Institute (Instituto de Ciencias Básicas) at the University of Veracruz (Universidad Veracruzana) in Xalapa, Veracruz, Mexico. She holds a BSc degrees in Agricultural Engineering (Polytechnic University of Madrid, Spain) and Food Science and Technology (Complutense University of Madrid, Spain) as well as a MSc and a PhD degree in Food Science (National Polytechnique Institute, Mexico City, Mexico). She is a member of the National Research System of Mexico (SNI), level 1 since 2017. Her research projects are focused on *ethno* biotechnology of foods, mainly in beneficial microorganisms, *Capsicum,* spp. and *Persea americana.* She has authored/co-authored 15 scientific papers up to date, 5 book chapters and, numerous scientific dissemination papers. Her presence is habitual in national and international conferences, and she has supervised thesis from bachelor, masters, and doctoral degrees. Finally, she is editor of the present book: Avocado: consumption and health.

Email: guiomel@gmail.com; gmelgar@uv.mx

Dr Alan Javier Hernández-Álvarez is currently a lecturer in Nutrition and Global Health at the School of Food Science and Nutrition, University of Leeds, UK. He holds a BSc degree in Biochemical Engineering and received a PhD in Food Science (National Polytechnic Institute, Mexico). During his PhD research, he developed his research work at the Food Phytochemistry Laboratory, Fat Institute, CSIC, Seville, Spain. He worked as a lecturer in Food Biochemistry, at the Biochemical Engineering Department at the National School of Biological Sciences (National Polytechnic Institute, Mexico, 2012-2013). Subsequently, he received a postdoctoral grant from CONACyT (in 2013) to continue his research at Agriculture and Agri-Food Canada, Saint Hyacinthe, Quebec, Canada. Afterwards, he was awarded with a Visiting Fellowships in Canadian Government Laboratories Award, NSERC (Natural Sciences and Engineering Research Council of Canada, 2014-2017). He collaborated with different Canadian Industries in an extensive variety of projects, some of these are InnoVactiV, Nature Zen, among others, as a research and scientific advisor. His research interests focus on molecular and biochemical mechanisms of prevention of bioactive food components, mainly proteins, bioactive peptides and phytochemicals, and their safety. He has authored/co-authored over 20 scientific papers and 5 book chapters and participated in more than 45 different national and international scientific conferences.

Email: A.J.HernandezAlvarez@leeds.ac.uk

ACKNOWLEDGMENTS: REVIEWERS

We, the editors, want to acknowledge the selflessly great work done by the researchers who reviewed the chapters that are part of this book, for their valuable comments and suggestions throughout the preparation of this book.

Dr. Cristina Torres Fuentes
Universitat Rovira i Virgili,
Nutrigenomics Research Group,
Biochemistry and Biotechnology Department,
Campus Sescelades, Tarragona, Spain
Email: cristina.torres@urv.cat

Dr. Janet Carrasco Castilla
Instituto Politécnico Nacional,
Centro de Estudios Científicos y Tecnológicos 17 León,
Guanajuato, Mexico
Email: jcarrascoc@ipn.mx

Dr. Matthew D. Campbell
School of Food Science and Nutrition,
University of Leeds, Leeds, United Kingdom.
Email: M.D.Campbell@leeds.ac.uk

Dr. Michael Zulyniak
School of Food Science and Nutrition,
University of Leeds, Leeds, United Kingdom
Email: M.A.Zulyniak@leeds.ac.uk

Dr. Martin Mondor
Agriculture and Agri-Food Canada. Science & Technology Branch,
St-Hyacinthe, Quebec, Canada
Email: Martin.Mondor@canada.ca

Dr. Alberto Sánchez Medina
Unidad de Servicios de Apoyo en Resolución Analítica,
Universidad Veracruzana. Veracruz, México
Email: asanchez@uv.mx

In: Avocado: Consumption and Health ISBN: 978-1-53618-204-0
Editors: M. G. Melgar Lalanne et al. © 2020 Nova Science Publishers, Inc.

Chapter 1

THERAPEUTIC AND NUTRACEUTICAL POTENTIAL OF PERSEA AMERICANA FRUITS AND THEIR BY-PRODUCTS: A REVIEW

***Ana F. Vinha*[1,2]*, Carla Sousa*[1] *and Carla Moutinho*[1,*]**
[1]UFP Energy, Environment and Health Unit (FP-ENAS),
University Fernando Pessoa, Porto, Portugal
[2]LAQV-REQUIMTE, Department of Chemical Sciences,
Faculty of Pharmacy, University of Porto, Porto, Portugal

ABSTRACT

Avocado (*Persea americana* Mill.), originated from Mexico and Central America, is the only commercially crucial edible fruit that belongs to the Lauraceae family. This fruit shows a growing value on the international market, not only because of its nutritional quality but also due to its medicinal uses and application in the cosmetic industry. Fruit constituents reveal that it abounds in various fatty acids, protein, vitamins, and minerals. Besides this, a lot of secondary metabolites have also been isolated from different parts of the avocado plant. The

[*] Corresponding Author's Email: acvinha@ufp.edu.pt.

predominant carotenoid is lutein, but other carotenoids are also described, including α-carotene, β-carotene, zeaxanthin, neoxanthin, and violaxanthin. It has been reported that these lipophilic carotenoids may have, among others, potential anti-carcinogenic effects. This fruit has been recognized for its health benefits, mainly due to the compounds present in the lipidic fraction, such as omega fatty acids, phytosterols, tocopherols, and squalene. Several studies have shown the benefits of avocado associated with a balanced diet, especially in reducing cholesterol and preventing cardiovascular diseases. The processed avocado pulp is an alternative to utilize fruits, which can be used in various value-added food products.

Tocopherols have also been identified not only in avocado pulp fruits but in their seeds. For these reasons, the recovery of health benefit bioactive compounds from fruit and its wastes is a research trend not only to help minimize the waste burden but also to meet the intensive demand from the public for bioactive compounds which are believed to have protective effects against chronic diseases.

Avocado pulp and seed contain variable oil content and is widely used in the pharmaceutical and cosmetic industries, and the production of commercial oils similar to olive oil. It contains fatty acids such as lignoceric acid, arachidic acid, margaric acid, behenic acid, gadolenic acid, docosadienoic, myristic, and eicosanoic acids, β-sitosterol, lecithin, minerals, and vitamins, showing its potential as an alternative dietary supplement.

This review is focused on the nutritional characterization, polyphenolic compounds recovery from avocado fruits, and their wastes (peel and seeds) as an excellent alternative for industry, especially for pulp processing or oil extraction, considering its composition and the benefits of its compounds. Furthermore, this crop can be used for exportation and oil extraction, application in processed products, or as a raw material in the pharmaceutical and cosmetic industries, generating high value-added products.

Keywords: *Persea americana* Mill, nutraceuticals, bioactive compounds, therapeutic properties, by-products, sustainability

1. Introduction

Avocado (*Persea americana* Mill) is a flowering plant belonging to Lauraceae family, genus Perseus, native from Mexico and Central

America, and which has recently been winning worldwide acceptance due to its benefits for human health (Araújo et al., 2018). Avocado is highly regarded today as an essential fruit for food and nutrition, and the remarkable development of this fruit has been recorded in every continent.

Avocados have been known for many years, dating back at least to 1519 when a Spanish soldier of fortune (Hernando Cortés) saw and reported the fruit tree in Colombia (Bill et al., 2014).

As already mentioned, Mexico, Central America, and northern South America are considered to be the centers of origin of avocados; however, avocados have been cultivated in a varying range of habitats, including Chile, Israel, Spain, South Africa, Peru, Kenya, USA, and the Dominican Republic, among others (Whiley et al., 2002). Archaeological and other evidence indicates that avocados were cultivated wherein very early times, possibly some 6000 years ago (Chen et al., 2009). They were not commercially exploited until the first years of the twentieth century when their potential for commercialization was developed by Californians. It was in this state that avocado growing became a recognizable industry. The crop has wide adaptability, and avocados are now grown in the whole world.

According to the Food and Agriculture Organization (FAO) (2010) reports, 13% of the avocados produced in Mexico are traded internationally. In Chile, South Africa, Israel, and Spain, the avocado production is mainly aimed at exporting the fruit to the overseas markets. The world avocado trade is focused on two major markets, more specifically, the European Union (EU), which imports 150,000–160,000 t per year, including imports from Spain, and the United States, which imports 140,000 t per year. According to De Bon (2010), Canada and Japan imported about 30,000 and 15,000 t, respectively, and the Asian, Middle East, and Eastern European markets imported less than 20,000 t per year. Peru, South Africa, and Israel are the major avocado exporters to the European markets (De Bon, 2010).

Commonly, this fruit is called avocado, characterized by its pear-shaped, egg-shaped, or spherical (Figure 1). Morphologically, avocado fruit is formed by pericarp that consists in three layers: the exocarp (1), which comprises the skin or rind, the fleshy mesocarp (2), that is the edible portion, and the endocarp (3), which is the thin inner layer next to the outer seed coat (Cummings and Schroeder, 1942; Cowan and Wolstenholme, 2003).

Concerning the cultivars most produced worldwide, Hass and Fuerte dominate the international market (Rodríguez-Carpena et al., 2011), but Shepard, Sharwil, Pinkerton, Reed, and Wurtz are also recognized (Rodríguez-Carpena et al., 2011). According to Dembitsky et al., (2011), there are about 500 varieties of avocado varieties, which differ mostly in the fruit shape and color. Although the Fuerte cultivar displays a green skin color after-ripening and is well known in the European markets, the Hass cultivar currently dominates the international trade market due to its longer shelf life and fruit size, and higher consumer acceptance due to its rich, nut-like flavor (FAO, 2010). Hass variety is the dominant and most popular type grown in the USA for export. The avocado export markets are segmented, according to cultivars. Hass cultivar differs significantly from other cultivars, including Fuerte, Reed, and Pinkerton, which have green skin (De Bon, 2010).

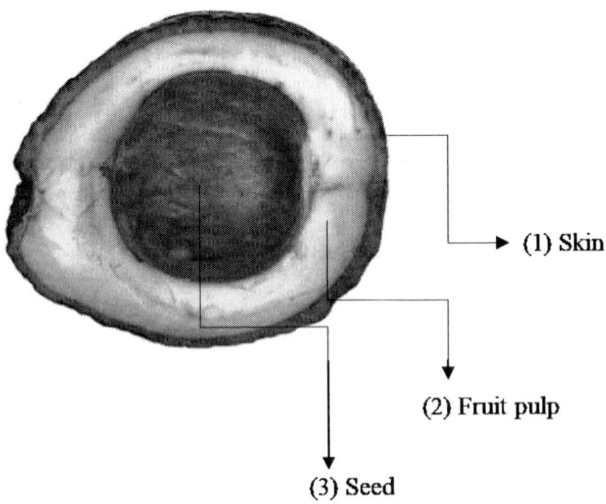

Figure 1. Morphological characteristics of avocado (*Persea americana* Mill.).

Despite their organoleptic differences, avocado varieties also present differences in their size and chemical composition. For example, Fuerte variety is larger and displays more weight than the Hass variety. Despite this, Hass variety presents a better ratio of an edible portion (pulp) than Fuerte variety. The proportion of fruit by-products (peel and seed) is higher in Fuerte variety than in the Hass variety. The pulp of Fuerte has lower fat content than the Hass variety (Rodríguez-Carpena et al., 2011).

Avocado is a well-known source of nutrients (fatty acids and proteins), bioactive compounds (carotenoids and phenolics), vitamins (E, B1, B2, and D) and minerals (calcium, iron) (Dembitsky et al., 2011; Tabeshpour et al., 2017). According to some researchers, avocado is the most nutritive among fruits and is regarded as the most significant contribution of the New World to the human diet (Rodríguez-Carpena et al., 2011; Melgar et al., 2018). The edible part (pulp) is rich in proteins (up to 4%) and fat (up to 30%), but low in carbohydrates (Tabeshpour et al., 2017). The fat is similar to olive oil in composition and is widely used in the preparation of pharmaceutical and cosmetics products (Tango et al., 2004; Ballmann and Müeller, 2008; Melgar et al., 2018. Therefore, the development of avocado production will help local people to improve their nutritional standards. Furthermore, avocado has the highest energy value (245 cal/100 g) of any fruit (Bill et al., 2014).

Currently, there are many patents with avocado uses in food, health, and cosmetic field, as a nutritional fruit juice, avocado wine, an herbal cream, or avocado-derived lipids for use in treating leukemia, among others (Araújo et al., 2018). Avocados are traditionally used for various purposes, from medicinal to culinary, dermo-cosmetic applications, and others (Nayak et al., 2008; Anaka et al., 2009; Kosińska et al., 2012). In folk medicine, avocados are also used as hypotensives, hypoglycemics, and antivirals and are applied for the treatment of ulcers, cardiac and dermatological diseases (Almeida et al., 1998; Adeboye et al., 1999; Ukwe and Nwafor, 2004; Antia et al., 2005; Owolabi et al., 2005; Raharjo et al., 2008). Analgesic and anti-inflammatory effects have also been reported (Adeyemi et al., 2002). Besides the antihypertensive and cardioprotective action, avocado also contributes to lipid-lowering, antidiabetic, anti-obesity, antithrombotic, and antiatherosclerotic effects (Tabeshpour et al., 2017). Avocados, and others such as dark chocolate, red wine, nuts, beans, are being promoted as anti-aging foods, due to their anti-oxidative and anti-inflammatory properties (Chrysohoou and Stefanadis, 2013).

Thus, in this chapter, it will refer to the nutritional, chemical, and biological properties of avocado, also highlighting its by-products (peel and seed) as industrial waste with high economic value for pharmaceutical and food industries.

2. Avocado Nutritional Composition

In the last decades, the evidence of the benefits of avocado for health has increased, and so, its consumption has increased, stimulating the investigation of its potential (Lu et al., 2009). Thus, in some countries, information on avocado and its importance to health has been highlighted. The independent Australian organization, "The Heart Foundation," certified the fruit as healthy food for the heart, while the Californian Avocado Commission encouraged efforts to promote avocado in the health area (Knight, 2002).

Moreover, in recent years, there has been a growing interest in functional food that can provide not only the basic nutritional and energetic requirements but also additional physiological benefits (Zhou et al., 2016; Duresa, 2017). Avocado is classified as a functional food due to its high nutritional value and proven beneficial effects for human health (Zhou et al., 2016; Duresa, 2017). Avocado is a nutrient-rich fruit, with the composition depending on edaphic and climatic conditions (e.g., ecotype, cultivar, degree of maturity, and growing conditions). Despite everything, avocado has been recognized for its high content of bioactive compounds present in the lipid fraction, such as fatty acids, phytosterols, tocopherols and squalene (Santos et al., 2014; Duarte et al., 2016). Avocados contain an oil rich in monounsaturated fatty acids (MUFA) in a water-based matrix, which enhances nutrient and phytochemical bioavailability and masks the taste and texture of the dietetic fiber (Dreher and Davenport,

2013). The avocado oil consists of 71% MUFA, 13% polyunsaturated fatty acids (PUFA), and 16% saturated fatty acids (SFA), which helps to promote healthy blood lipid profiles (Dreher and Davenport, 2013). According to Lu et al., (2009), pulp fruit is rich in lipids, which vary from 3% to 30% of its fresh weight, taking into account that oleic, palmitic, linoleic, and palmitoleic acids are the predominant fatty acids, as well as stearic acid in trace amounts. Santos et al. (2014) reported that oleic fatty acid represented more than half of the total fatty acids of this raw material, together with substantial amounts of unsaturated linoleic and palmitoleic acids. Lately, it has been discovered a new lipid derived from the avocado fruit, avocatin B, with cytotoxic activity in acute myeloid leukemia and to inhibit fatty acid oxidation and decrease NADPH levels, resulting in ROS-dependent leukemia cell death (Araújo et al., 2018). In contrast to other vegetable oils, avocado oil is characterized by having high levels of oleic and palmitoleic acids, low linoleic acid, and relatively high levels of palmitic and stearic acids. This fatty acid composition is influenced by the cultivars, maturity stage, the anatomical region of the fruit, and geographic location for plant growth (Contreras-Gutiérrez et al., 2013; Duarte et al., 2016; Corzzini et al., 2016; Forero-Doria et al., 2017). Forero-Doria et al., (2017) studied the oxidative stability of Chilean avocado oil and assessed the relationship between the antioxidant capacity of the oil and its protection against the induced thermal oxidation, considering their different fatty acid composition. The results showed a high content of monounsaturated fatty acid, with a percentage of 69.4%, and the amount of polyunsaturated and saturated fatty acids was 16.6 and 14%, respectively. The composition of MUFA included the presence of oleic (65.3%) and palmitoleic (3.9%) fatty acids, whereas the PUFA content comprised the following fatty acids: linoleic (15.2%) > linolenic (1.3%) > araquidonic (0.1%). Additionally, the SFA content corresponded to the presence of stearic and palmitic acids in 0.6 and 13.4%, respectively. According to Bill et al., (2014), the fatty acid composition in the avocado mesocarp of different avocado cultivars is as follows: oleic acid (86 mg g-1 oil), palmitic acid (32 mg g-1 oil), linolenic acid (19 mg g-1 oil), and palmitoleic acid (14 mg g-1 oil). Dávila et al., (2017) reported that the major fatty acids of the Hass avocado oil were oleic acid (C18:1, 50.96%), palmitic acid (C16:0, 24.74%,), linoleic acid (C18:2, 15.07%), palmitoleic acid (C16:1, 5.97%) and linolenic acid (C16:3, 1.66%). Other fatty acids such as myristic (C14:0, 0.05%), margaric (C17:0, 0.03%), margaroleic (C17:1, 0.08%), estearic (C18:0, 0.84%), arachidic (C20:0, 0.14%), gadoleic (C20:1, 0.27%), behenic (C22:0, 0.27%), lignoceric (C24:0, 0.06%) and other isomers trans of C18 were present only at low concentration. However, the content of fatty acids also depends on the drying and extraction methods of oil. According to Krumreich et al., (2018), fatty acids of avocado oil were better preserved when the pulp was dried at 60°C and using mechanical pressing. Palmitic (~21.0%), palmitoleic (~7.0%), linoleic (~11.0%), and tricosanoic (~2.5%) acids and lower relative proportions of linolenic acids (~0.6%) and behenic acids (~0.76%) were identified in avocado oil in addition to oleic acid (~58.0%).

Considering other constituents, avocado fruit carbohydrates are composed of about 80% dietary fiber (70% insoluble and 30% soluble fiber) (Marlett and Cheung, 1997). Avocados contain levels between 2.0 g and 4.6 g of dietary fiber per 30 g and one-half fruit, respectively (USDA, 2011). Thus, moderate avocado consumption can help to achieve an adequate intake of 14 g dietary fiber per 1000 kcal as about one-third this fiber level can be met by consuming one-half an avocado (Dreher and Davenport, 2013). Avocados contain a low content of sugar. One-half of an avocado contains 0.2 g sugar (e.g., sucrose, glucose, and fructose). The primary sugar found in avocados is a unique seven-carbon sugar (D-mannoheptulose) Moreover, its reduced form, perseitol, that does not behave nutritionally as regular sugar and is more of a unique phytochemical to avocados (Meyer and Terry, 2008). Roth (2009) suggested that D-mannoheptulose may support blood glucose control and weight management.

Salgado et al. (2008) reported that 70% of total fruit weight corresponds to the pulp fruit with an average of 6.94 g of carbohydrates, 17.34 g of fat, 2.08 g of proteins, 2.72 g of fibers (in 100 g of fresh pulp). It is stated here that the fat content is, undoubtedly, the nutritional fraction with the highest expression in avocados. Moreover, according to USDA databases (2011) and Bes-Rastrollo et al., (2008), avocados are a medium energy-dense fruit, just because nearly 80% of avocado fruit pulp consists of water (‘‘72%) and dietary fiber (‘‘6.8%) and has been shown to have similar effects on weight control as low-fat fruits and vegetables.

Avocado industrial processing generates large quantities of agro-industrial by-products (peel and seeds), ranging from 18% to 30% of dry fruit weight (Wang et al., 2010; López-Cobo et al., 2016). Thus, it is interesting to reuse these by-products both to reduce their negative impact on the environment and to add value to them, since they are sources of critical phytochemical compounds. A by-product of avocado is its peels. Forty percent of the livestock herders' budgets, especially in zero-grazing, go-to food supplements. With avocado peel, rich in oil (7%), minerals, and other acids, it is possible to cut much of the feed costs. Research has shown that the peels have become part and parcel of commercially available feed (Gómez et al., 2014; Araújo et al., 2018). According to some authors, the composition of avocado by-products varies between cultivars, climatic conditions, genetics, and others (Gómez et al., 2014; Araújo et al., 2018). The predominant constituents of avocado by-products (peel and seeds) are carbohydrates, such as fibres, hemicellulose, and starch (seed), which can be a potential source of energy (bioethanol) and produce other added-value compounds (Ruiz et al., 2013). Thus, avocado consumption, increase colonic mucin 3 and 4 gene expression as well as the number of goblet cells (mucin-producing cells) (Paturi et al., 2017). Avocado fruit seeds are a potential alternative fiber source. This makes them a candidate for study as part of the current boom in identifying applications for vegetable by-products, such as dietary fiber. Moreover, avocado by-products represent a prospective source of phytochemical compounds with biological properties, may containing more polyphenols than other fruits recognized for their antioxidant activity (Ruiz et al., 2013).

2.1. Avocado Phytochemical Profile

Natural antioxidants, notable polyphenols, have the potential to be used in therapeutic fields because of their numerous benefits, like reducing the risk of inflammatory diseases and preventing lipid oxidation (Haque and Mahadeva, 2012; Segovia et al., 2016). They are also known to have effects against some human neurodegenerative diseases (Ghasemzadeh and Ghasemzadeh, 2011; Kaur et al., 2016; Wang et al., 2016). Flavonoids and phenolics acids are the most prestigious groups of secondary metabolites and bioactive compounds in plants (Ghasemzadeh and Ghasemzadeh, 2011). According to its origin, phenolic acids can be classified as derived from benzoic acid or cinnamic acid (Hurtado-Fernández et al., 2013; Hurtado-Fernández et al., 2014; Hurtado-Fernández et al., 2015).

Due to their structural similarities with several key biological effectors and signal molecules, phenols can participate in the induction/repression of gene expression or activation/deactivation of proteins, enzymes, and transcription factors of critical metabolic pathways (Tesfay et al., 2011). Phenols can also play a role in modulating cellular homeostasis because of their physicochemical properties (Tesfay et al., 2011). Varied biological activities of phenolic acids were reported: increases bile secretion, exhibits hepatoprotective effect, reduces blood lipid levels and shows antiviral, fungicidal and antimicrobial activities and larvicidal effects (Ghasemzadeh and Ghasemzadeh, 2011; Dabas et al., 2013; Barbosa-Martin et al., 2016; Kaur et al., 2016; Yamassaki et al., 2017). Epidemiologic findings revealed that polyphenols have cardioprotective effects, which include inhibition of platelet aggregation and vascular relaxation through the production of nitric oxide (NO), which decreases low-density lipoprotein (LDL) oxidation and prevent atherosclerosis (Haque and Mahadeva, 2012). Phenolic compounds participate in the prevention of DNA adduct formation and enhanced carcinogen elimination. Additionally, phenolic plant compounds are capable of inhibiting inflammation and tumor promotion by deactivating a range of pro-oxidative enzymes (Calderón-Oliver et al., 2016; Rani et al., 2016; Wang et al., 2016).

Avocados contain a moderate level of phenolic compounds contributing 60 mg and 140 mg gallic acid equivalents (GAE) per 30 g and one-half fruit, respectively (Dreher and Davenport, 2013). It has been reported that the antioxidant capacity of fruits and vegetables has been correlated to their total phenolic content and composition (Nour et al., 2013; Calderón-Oliver et al., 2016). The avocado also has a total antioxidant capacity of 600 μmol Trolox Equivalent (TE) per 30 g or 1350 μmol TE per one-half fruit. Also, avocados have the highest fruit lipophilic antioxidant capacity, which may be one factor in helping to reduce serum lipid peroxidation and promoting vascular health (Dreher and Davenport, 2013).

A research performed by Golukcu and Ozdemir (2010) on Bacon, Zutano, Fuerte, and Hass avocado mesocarp part indicated that protocatechuic acid, caffeic acid, (-)-epicatechin, and rutin were the main phenolic compounds. A few other examples of phenolics found were p-hydroxybenzoic, protocatechuic, β-resorcyclic, γ-resorcyclic, α-resorcyclic, gallic, isovanillic, vanillic, syringic, o-coumaric, m-coumaric, p-coumaric, ferulic, and sinapic acids.

The phenolic profiles of six varieties of avocado (Hass, Bacon, Fuerte, Pinkerton, Rincon, and Orotawa) grown in Sicily (Italy) at two different ripening stage (unripe: just harvested; ripe: ready for consumption), were investigated by ultra-high-performance liquid chromatography-heated electrospray-mass spectrometry (Stefano et al., 2017). Nineteen individual phenolic compounds were established in the studied extracts, including gallic acid, sinapinic acid, vanillin, p-coumaric acid, and gentisic acid were presented only in ripe fruits. In contrast, epicatechin decreased with fruit ripening, whereas protocatechuic acid, 4-hydroxybenzoic acid, chlorogenic acid, and benzoic acid were moderately stable with fruit ripening. Caffeic acid, ferulic acid, taxifolin, rutin, trans-cinnamic acid, isorhamnetin were also presented in ripe fruit from the six avocado cultivars, mainly in Hass variety. Poncirin and narirutin were only founded in Pinkerton cultivar. It has also been shown that the different avocado cultivars presented different phenolic total quantities, and it was found that Orotawa may provide the highest nutritional and health contribution to the human diet (Stefano et al.,

2017). Rodríguez-Carpena et al. (2011) determined the total phenolics content, phenolics profile, and antioxidant potential of different extracts (ethyl acetate, acetone, or methanol) from the peel, pulp, and seed of the two most typical avocado varieties (Hass and Fuerte). The authors observed that peels and seeds were rich in catechins, procyanidins, and hydroxycinnamic acids, whereas the pulp was particularly rich in hydroxybenzoic and hydroxycinnamic acids and procyanidins. The total phenolic content and antioxidant potential of avocado phenolics were affected by the extracting solvent and avocado variety.

The avocado processing industry generates significant amounts of waste (peel and seed) that can thus potentially be processed to yield value-added products with promising and significant applications in pharmaceutical industries (Segovia et al., 2016; Dávila et al., 2017).

Avocado seeds and leaves, as aqueous and hydroalcoholic extracts, are utilized in folk medicine because of their anti-viral, anti-inflammatory, anti-diabetic, hypoglycemic, and antioxidant properties - these beneficial biological properties may be related to the presence of polyphenolic compounds (Brai et al., 2014; Calderón-Oliver et al., 2016). Several studies of avocado seed and peel have described the presence of phenolic compounds belonging to five different groups: catechins, hydroxybenzoic acids, hydroxycinnamic acids, flavonols, and procyanidins, with the latter group being the most abundant one (Kosińska et al., 2012).

Some authors have reported the presence of phenolic acids, flavonoids, and procyanidins in avocado leaves (Kumar et al., 2017; Yamassaki et al., 2017; Tremocoldi et al., 2018). From the chemical point of view, avocado seeds contain high levels of phenolic compounds (more than 50%), such as hydroxybenzoic acids, hydroxycinnamic acids, flavanols (such as catechin, epicatechin), flavonols (such as rutin), and procyanidins (Pahua-Ramos et al., 2012; Soldera-Silva et al., 2018). According to Tremocoldi et al., (2018), the main compounds with significant contribution to the antioxidant activity were procyanidin B2 and epicatechin in the peel and trans-5-O-caffeoyl-D-quinic acid, procyanidin B1, catechin, and epicatechin in the seed of avocado varieties Hass and Fuerte.

Hepatoprotective activity of aqueous leaf extract of *P. americana* was demonstrated against carbon tetrachloride induced rat liver damage, and this effect was attributed to its phytochemicals constituent (Brai et al., 2014). Antidiabetic activity of hydroalcoholic extract of the leaves was confirmed (Lima et al., 2012), and antidiarrhoeal activeness was recognized in chloroform-methanol leaf extract of P. americana (Christian et al., 2014). Avocado seeds maybe use as a source of polyphenols with promising anthelmintic applications (Soldera-Silva et al., 2018).

Avocado by-products also presented a very distinct phenolic profile, presenting higher concentration in peels (227.9 mg/g of extract for total phenolic content), mainly in (epi)catechin derivatives (175 mg/g of extract), followed by chlorogenic derivatives (42.9 mg/g of extract) (Melgar et al., 2018). The main family of phenolic compounds found in avocado peels and seeds was (epi)catechin derivatives. The second major family of compounds found in avocado peels samples was flavonoids, mainly quercetin derivatives (Barbosa-Martin et al., 2016; Dávila et al., 2017; Saavedra et al., 2017). Moreover, Shehata and Soltan (2013) likewise evaluated the effect of avocado (fruit and seed) on diet-induced hypercholesterolemia in rats. The results showed that avocado seed content was the highest in phenolic compounds (285.43 mg/100 g GAE). The lowering effect of avocado seed on lipid profiles in serum and liver were investigated in high-fat diet rats for six weeks. The results showed that both the fruit and the seed, particularly the later, significantly decreased the serum levels of total cholesterol, triglycerides, and low-density lipoprotein cholesterol (LDL-c).

Vinha et al. (2013) have demonstrated that the pulp of the Algarvian avocado is rich in phenolic compounds (410.2±69.0 mg/100g) and flavonoids (21.9±1.0 mg/100 g). It was concluded in the same studies that the levels of the peel were superior to the pulp with 679.0±117.0 mg/100 g of total phenolics and 44.3±3.1 mg/100 g of flavonoids. The seed, in turn, was the part of the fruit with the highest total phenolic content (704.0±130.0 mg/100 g) and flavonoids (47.97±2.69 mg/100 g). These phenolics and flavonoids are known to prevent the risk of developing certain diseases related to age, such as premature aging, cancer, and heart disease.

Kumar et al., (2017) evaluated the anti-inflammatory activity and antigenotoxic effect of hydroalcoholic leaf extract of *Persea americana* in albino Wistar rats against the whole body X-ray irradiation. *P. americana* at the optimum dose of 100 mg/kg body weight protected animals from clastogenic effects caused by X-rays and displayed anti-inflammatory properties. This protection was attributed to free radical scavenging activity of *P. americana* by maintaining cytoplasm antioxidant levels in cells, subsequently decreasing oxidative stress and inhibiting DNA damage, where flavonoids and polyphenolic compounds played a lead role in neutralizing the free radicals formed during irradiation.

Figueroa et al., (2018) determined the phenolic profile and other polar compounds of avocado seed and seed coat using accelerated solvent extraction (ASE) and liquid chromatography coupled to Ultra-High-Definition Accurate-Mass Q-TOF. In this work, 84 compounds were classified within eight subclass groups; among these, 45 phenolic compounds were identified for first time in the avocado seed. Condensed tannins, phenolic acids, and flavonoids were the most representative groups in both samples. Respect to phenolic acids, different sub-classes of these compounds has been characterized in both sample matrix, mainly hydroxybenzoic acids, hydroxyphenyl acetic acids, and hydroxycinnamic acids. The same authors (2018b) identified the phenolic compounds present in an extract of avocado peel obtained by a green extraction technique. Accelerated solvent extraction was performed using water and ethanol as extraction solvents. Liquid chromatography, coupled to ultra-high-definition accurate-mass spectrometry, was used in order to identify the bioactive compounds. A total of sixty-one compounds belonging to eleven families were identified. Procyanidins, flavonols, hydroxybenzoic and hydroxycinnamic acids were the most common compounds.

Besides polar phytochemicals, avocado also contains other bioactive compounds, nonpolar ones. The principal carotenoids of avocado belong to a subclass identified as xanthophylls, which are oxygen-containing fat-soluble antioxidants. Xanthophylls, such as lutein, β-cryptoxanthin, antheraxanthin, neoxanthin, violaxanthin, and zeaxanthin, are more hydrophilic than carotenes (which includes β-carotene and α-carotene), so they have a much lower tendency for pro-oxidant activity (Ortiz-Avila et al., 2013; Duarte et al., 2016).

Because carotenoids are hydrophobic, their absorption depends upon efficient release from the food matrix and subsequent solubilization by bile acids and digestive enzymes, culminating in their incorporation into micelles (Kopec and Failla, 2018). Avocado fruit has a unique unsaturated oil and water matrix naturally designed to enhance carotenoid absorption. Some studies have demonstrated that adding avocado fruit can significantly enhance (by 2-5 times) carotenoid absorption from salad and salsa, which is attributed mostly to the lipids present in avocado. In this manner, adding avocado fruit to carotenoid-containing foods can facilitate carotenoid absorption, and consequently aids to its lipid-decreasing function (Fulgoni et al., 2013). Lutein and zeaxanthin are selectively taken up into the macula of the eye, decreasing age-related eye dysfunction. Avocado fruits comprise 185 μg of lutein/zeaxanthin per one-half fruit, which is expected to be more highly bioavailable than most other fruit and vegetable sources (Dreher and Davenport, 2013). An *in vitro* study performed by Lu et al. (2009) showed that the avocado pulp, specifically lutein, and other carotenoids (as active compounds), possess growth inhibitory effects against cancer. The authors observed that the color of avocado pulp varies from dark green just under the skin to pale green in the middle section of the flesh to yellow near the seed. The total carotenoid concentrations were found to be greatest in the dark green pulp close to peel. It was also possible to conclude that in Hass avocados, xanthophylls prevail over the carotenes, contributing about 90% of the total carotenoids. Carotenoids, including lutein (2.93 μg g-1), zeaxanthin (0.11 μg g-1), β-cryptoxanthin (0.25 μg g-1), β-carotene (0.60 μg g-1), and α-carotene (0.25 μg g-1), were reported in the pulp of ripened Hass - the actual level of carotenoids in avocado was higher than the reported data because an average recovery from extraction, saponification and HPLC was 77%. Subsequently, the following carotenoids were identified: all-trans neoxanthin, all-trans violaxanthin, all-trans neochrome, 9′-cis-neoxanthin, lutein-5, six epoxide, and chrysanthemaxanthin. Vinha et al., (2013) also reported a total content of carotenoids in avocado pulp significantly inferior to in fruit peel (0.815±0.201 mg/100 g and 2.585±0.117 mg/100 g, respectively). Condelli et al. (2015) reported the carotenoid content of 10.3-14.4 mg kg^{-1}, while Krumreich et al., (2018) obtained a carotenoid content of 4.63 mg kg−1 using avocados of Breda variety.

Avocados are the richest known fruit source of phytosterols with about 26 mg and 57 mg per 30 g and one-half fruit, respectively (Dreher and Davenport, 2013). Some authors suggest that increased consumption of foods containing plant sterols has the potential to reduce the incidence of coronary heart disease and thus reduce costs associated with treating that disease in a significant way (Gyles et al., 2010; Kumar et al., 2017). Avocado contains substantial amounts of bioactive compounds such as phytosterols, especially in the lipid fraction, and the main representative is β-sitosterol (Ulbricht, 2016). Santos et al. (2014) investigated the influence of different procedures of pulp drying and oil extraction methods on the concentrations of several phytosterols in avocado oil. Pulp portions of Fortune variety avocados were dried either by lyophilization or under circulating air at 40 or 70°C, and the oil was obtained either by cold pressing or with Soxhlet extraction using petroleum ether. The concentration range found in this study was: β-sitosterol (86.6 - 91.2%), higher than campesterol (9.7 - 13.3%), higher than stigmasterol (0.04 - 0.06%). The highest yield (57%) was found for Soxhlet extraction after lyophilization.

The health effects of sterols and stanols have been the subject of several studies. The meta-analysis performed by Ras et al. (2013) reported that intake of 0.3-3.2 g of plant sterols reduces by 8.5% LDL-cholesterol, which is a well-known cardiovascular risk factor. The β-sitosterol in avocados has a unique effect on immunity, contributing to the treatment of diseases such as cancer, human immunodeficiency virus (HIV), and infections (Haque and Mahadeva, 2012; Dreher and Davenport, 2013). Studies have also shown that β-sitosterol activity is an aid in weight loss by reducing compulsive eating binge and fat accumulation in the abdominal region (Uchenna et al., 2017).

Furthermore, the daily consumption of phytosterols and phytostanols decreases the plasmatic levels of cholesterol, leading to the prevention of cardiovascular diseases and cancer progress (Tabeshpour et al., 2017).

An *in vitro* study showed that the avocado pulp, rich in tocopherols, lutein, and other carotenoids (as active constituents), possess growth inhibitory effects against cancer (Lu et al., 2009).

3. Biological Properties

There is already evidence that the ingestion of fruits confers protection against human chronic diseases, neurological disorders, and some types of cancer (Larijani et al., 2014; Reczek and Chandel, 2018). These properties are assigned to the presence of bioactive compounds in fruits (edible and not edible portion) (Krumreich et al., 2018). During the last decades, ample evidence of the benefits of avocado on health has been gathered. This promoted their consumption, also stimulating the research about their pharmacological potential. Moreover, avocado waste has gained more interest, not only because it is a cheap raw material, as well as its richness in biological compounds (Krumreich et al., 2018). Table 1 describes some of the biological properties described in the pulp and by-products of avocado fruit.

3.1. Cardiovascular Benefits

One of the essential mechanisms for the development of atherosclerosis is the oxidation of cholesterol-rich plasma lipoproteins. Oxidation of lipids increases their atherogenicity, facilitating their penetration into the arterial wall (Tiwari, 2004). The oxidation of lipoproteins is mainly due to the lipid peroxidation reaction initiated by free radicals (Tiwari, 2004). The main risks for the appearance of atherosclerosis are high levels of total cholesterol, low-density lipoproteins (LDL), triglycerides, and apolipoproteins B and C-III, as well as low levels of high-density lipoproteins (HDL) cholesterol and apolipoproteins A-I (Tiwari, 2004). According to a study performed by Pérez-Méndez and Hernández (2007), after four weeks of a diet rich in avocado, there was a significant reduction in triglycerides, with the maintenance of low blood lipid levels and glycemic index control. Since avocado is characterized by its high content of unsaturated fatty acids, the ingestion of this fruit brings benefits in maintaining the health of the heart and the circulatory system, since the replacement of saturated fats with unsaturated fats promotes the reduction of triglyceride levels, cholesterol, and LDL-cholesterol, and may even increase the HDL-cholesterol fraction, it is therefore recommended to introduce the fruit into the diet of people with cardiovascular disorders or diabetes (Tango et al., 2004). Pahua-Ramos et al., (2012) established that avocado seed flour reduced cholesterol in a model of hypercholesterolemic mice, being this effect attributed to the phenolic content, antioxidant activity, and the dietary and crude fiber contents of the seeds. A study performed by Márquez-Ramírez et al., (2018) in hypertensive rats have proved that avocado oil mimics the effects of losartan, an angiotensin-II antagonist, decreasing blood pressure, enhancing renal vasodilation, and counteracting mitochondrial dysfunction and excessive reactive oxygen species production.

Since avocado is a nutrient-dense source of monounsaturated fatty acids, it can be used to replace saturated fatty acids in a diet to lower low-density lipoprotein cholesterol. There are several studies about the effect of avocado consumption on cardiovascular disease risk factors. As avocado is associated with cardiovascular benefits, Mahmassani et al., (2018) studied the effect of the consumption of this fruit on cardiovascular disease risk with the use of a systematic review and meta-analysis. These authors conclude that although avocado intake resulted in no difference in serum total cholesterol, LDL-cholesterol, and triglycerides concentrations, it increased serum HDL-cholesterol concentrations.

Table 1. The most important bioactive compounds of avocado

Categories	Bioactive constituents	Biological activities	Avocado layer	References
Fatty Acids	oleic, palmitic, linoleic, stearic and palmitoleic acids; *acetogenins, (persenone-C and persediene)	decrease blood pressure, enhance renal vasodilation, and counteract mitochondrial dysfunction and excessive ROS production; prevent factor of metabolic syndrome; decrease high sensitivity C-reactive protein levels; *reduce inflammatory processes	pulp/ oil	Márquez-Ramírez et al. (2018) Dávilaet al. (2017) Forero-Doriaet al. (2017) Tabeshpouret al. (2017) Corzziniet al. (2016) Duarte et al. (2016) Rodríguez-Sánchezet al., (2015) Bill et al. (2014) Carvajal-Zarrabalet al., (2014) Santos et al. (2014) Contreras-Gutiérrezet al., (2013) Rodríguez-Sánchezet al., (2013) Lu et al. (2009) Tango et al. (2004)
Phytosterols	β-sitosterol	contribute to the treatment of diseases such as cancer, HIV and infections; reduce compulsive eating binge and fat accumulation in the abdominal region; decrease plasmatic levels of cholesterol, prevent cardiovascular diseases and reduce the incidence of coronary heart disease; prevent cancer progress; potential neuroprotective effect	pulp/seed	Kumar et al. (2017) Tabeshpouret al. (2017) Uchennaet al. (2017) Ulbricht (2016) Padmanabhan and Arumugam (2015) Dreher and Davenport (2013) Haque and Mahadeva (2012) Gyleset al. (2010)
Other lipids	avocatin B (avocadene and avocadyne)	cytotoxic activity in acute myeloid leukemia; inhibit fatty acid oxidation; decrease NADPH levels	pulp/seed	Ahmed et al. (2018) Araújo et al. (2018) Lee et al. (2015)
Vitamins	tocopherols	growth inhibitory effects against cancer	pulp	Thongchai (2014) Lu et al. (2009)
Carbohydrates	D-mannoheptulose and its reduced form (perseitol)	blood glucose control and weight management	pulp	Roth (2009) Meyer and Terry (2008)

Table 1. (Continued)

Categories	Bioactive constituents	Biological activities	Avocado layer	References
Carotenoids	Xanthophylls (lutein, zeaxanthin, β-cryptoxanthin), carotenes (β-carotene and α-carotene)	inhibitory effects against cancer	pulp	Dreher and Davenport (2013) Lu et al. (2009)
Polyphenolic compounds	protocatechuic, caffeic, *p*-hydroxybenzoic, hydroxycinnamic, protocatechuic, β-resorcyclic, γ-resorcyclic, α-resorcyclic, gallic, isovanillic, vanillic, syringic, o-coumaric, m-coumaric, *p*-coumaric, ferulic, and sinapic acids, epicatechin and rutin procyanidins, hydroxybenzoic and hydroxycinnamic acids, flavonoids: flavonols (rutin), flavanols (catechin, epicatechin), (epi)catechin derivatives, chlorogenic derivatives, and quercetin derivatives	decrease serum levels of total cholesterol, triglycerides, and LDL; prevent the risk of developing certain diseases related to age, such as premature aging, cancer, and heart disease; potential neuroprotective effect	pulp	Padmanabhan and Arumugam (2015) Shehata and Soltan (2013) Vinha et al. (2013) Rodríguez-Carpena et al.,(2011) Golukcu and Ozdemir (2010)
		anti-viral, anti-inflammatory, anti-diabetic, hypoglycemic, and antioxidant activities; decrease serum levels of total cholesterol, triglycerides, and LDL; anthelmintic applications; antigenotoxic effect against X-ray irradiation	Avocado by-products (peel/seed/leaves)	Melgar et al. (2018) Soldera-Silva et al. (2018) Tremocoldiet al. (2018) Dávila et al. (2017) Kumar et al. (2017) Saavedra et al. (2017) Yamassaki et al. (2017) Barbosa-Martin et al. (2016) Calderón-Oliver et al., (2016) Brai et al. (2014) Shehata and Soltan (2013) Kosińska et al. (2012) Pahua-Ramos et al. (2012) Rodríguez-Carpena et al.,(2011)
Peptides	glutathione	Reduction of the risk of the development of larynx, pharynx and oral cavity cancers		Castillo-Juarez et al., (2009) Flagg et al. (1994)

The objective of the study of Carvajal-Zarrabal et al. (2014) was to evaluate the effects of avocado oil administration on biochemical markers of cardiovascular risk profile in rats with metabolic changes induced by sucrose ingestion. This study proved that avocado oil is a potential preventive factor of metabolic syndrome since it reduces triglycerides (TG), very low-density lipoprotein (VLDL) and low-density lipoprotein (LDL) levels, without affecting high-density lipoproteins (HDL) values. Avocado oil, extracted by centrifugation or solvent, decreased high sensitivity C-reactive protein levels, indicating that inflammatory processes were partially reversed, suggesting that the supplementation with this oil has a positive health outcome since it reduces inflammatory processes and produces positive changes in the biochemical indicators related to the development of the metabolic syndrome. Metabolic syndrome is an aggregation of risk factors like dyslipidemia, hypertension, high blood glucose, and obesity that result in an increased risk of type 2 diabetes mellitus and cardiovascular diseases, which are among leading causes of death in the world (Tabeshpour et al., 2017). Another meta-analysis with healthy adults with an average body mass index performed by Peou et al., (2016) obtained similar results: avocado consumption significantly reduced total cholesterol, LDL, and TG, decreasing HDL nonsignificantly. The aim of the work developed by Furlan et al., (2017) was to evaluate the impact of exchanging butter by Hass avocado-oil on postprandial metabolic parameters in healthy overweight volunteers consuming a hypercaloric-hyperlipidemic breakfast. This study proves that the replacement of butter by Hass avocado-oil reduced postprandial inflammation and postprandial glycemia. The results of these authors also contributed to consider Hass avocado-oil a potential atherosclerosis-protective fat.

Inclusion of one avocado per day as part of a moderate-fat and cholesterol-lowering diet of a healthy overweight adult with LDL cholesterol 105 - 194 mg/dL for males and 98 - 190 mg/dL for females and normal or well-controlled blood pressure has additional LDL and non-HDL lowering effects (Wang et al., 2015), in agreement with the results obtained by Carvajal-Zarrabal et al., (2014). As the results of other studies, this demonstrates that avocados have beneficial effects on cardio-metabolic risk factors that extend beyond their heart-healthy fatty acid profile. So, substitute dietary fats, especially saturated ones, by avocados should be one of the health recommendations. Previous studies suggest that the daily consumption of phytosterols and phytostanols, important bioactive phytochemicals of avocado, decreases the serum levels of cholesterol, leading to the prevention of cardiovascular disease and also possible protection in the number of cancer development (Awad and Fink, 2000; Piironen et al., 2000; Kritchevsky and Chen, 2005).

Avocado fruit is a rich source of health-related lipophilic phytochemicals such as monounsaturated fatty acids, tocopherols, carotenes, acetogenins, and sterols. Lipophilic antioxidant capacity of avocado pulp is higher than hydrophilic capacity (Rodríguez-Sánchez et al., 2013). Acetogenins, namely Persenone-C and Persediene, were identified as a novel lipophilic antioxidant present in the avocado pulp. In the presence of emulsifiers, acetogenins may serve as lipophilic food antioxidants (Rodríguez-Sánchez et al., 2013). In effect, the avocado pulp is a good source of fatty acids related to platelet aggregability reduction, preventing thrombus formation responsibly for ischaemic diseases. Rodríguez-Sánchez et al. (2015) identified Persenone-C as the most potent antiplatelet acetogenin *in vitro. In vivo*, Persenone A demonstrated potential protection against arterial thrombosis, as coagulation times increased, and thrombus formation was attenuated. Besides the fruit and avocado oil, its by-products also have important cardiovascular properties. Although the seed represents a considerable percentage of the total fruit, currently, it still represents an under-utilized resource and a waste issue. Avocado seeds may improve hypercholesterolemia and be useful in the treatment of hypertension, inflammatory conditions, and diabetes (Dabas et al., 2013).

3.2. Anticancer Activity

Extensive research over the last decades indicates that excessive accumulation of reactive oxygen species (ROS) can promote cancer.

Cancer cells produce increased numbers of ROS and depend on an antioxidant defense system. Besides, ROS promotes cancer cell survival. Reactive oxygen species are produced as an intracellular byproduct by mitochondria and other cellular elements and exogenously by pollutants, tobacco, smoke, drugs, xenobiotics, and radiation (Prasad et al., 2017; Reczek and Chandel, 2018). It is believed that ROS are related to the genesis of tumors at different levels, and can promote the development of cancer through different mechanisms: alterations in cellular proliferation; the decreased affinity of the polymerase enzyme; inactivation of enzymes responsible for the degradation of carcinogens; binding of end products of lipid peroxidation to the DNA molecule, creating mutagenic lesions; direct DNA attack (Min and Ebeler, 2008).

Several studies showed that phytochemicals extracted from the avocado fruit selectively induce cell cycle arrest, inhibit growth, and induce apoptosis in precancerous and cancer cell lines. Rajkumar et al. (2011) reported that some phytochemicals extracted with 50% methanol from avocado fruits help in the proliferation of human lymphocyte cells and decrease chromosomal aberrations induced by cyclophosphamide. The authors suggested that phytochemicals from the avocado fruit can be utilized for making active chemoprotective ingredients for lowering the side effect of chemotherapy in cancer therapy. The phytochemicals present in the avocado extract also inhibits the growth of oral cancer and pre-cancer cells and induces apoptosis, without affecting the healthy cells (D'Ambrosio et al., 2011).

A diet rich in antioxidants, such as carotenoids (Nagarajan et al., 2017; Eggersdorfer and Wyss, 2018) and phenolic compounds (Zhang et al., 2018; Jimenez-Garcia et al., 2018), may prevent carcinogenesis and studies are describing the beneficial effect of consumption of fruits source of these compounds in reducing the risk of various types of cancer. Studies conducted *in vitro* and *in vivo* indicate that avocado should be included in the diet for its cancer-preventive benefits. This fruit contains a large number of phytochemical compounds, among which, carotenoids, terpenoids, D-mannoheptulose, phenols, and glutathione, described as compounds with anticancer properties (Ding et al., 2009). Another *in vitro* study showed that tocopherols, lutein, and other carotenoids, existing in the avocado pulp, possess growth inhibitory effects against cancer (Lu et al., 2009). Regular consumption of phytosterols and phytostanols, bioactive compounds of avocado, has a potential protective effect in cancer development (Awad and Fink, 2000; Piironen et al., 2000; Kritchevsky and Chen, 2005).

The effectiveness of avocado phytochemicals in cancer treatment has been investigated only on prostate, breast, oral, esophageal, and colon cancers (Larijani et al., 2014). Avocado contains soluble lipid components such as tocopherols and carotenoids with anti-cancer activities, and it can significantly reduce the risk of androgen-dependent and -independent prostate cancer.

Bonilla-Porras and coworkers (2014) demonstrated that endocarp, seed, whole seed, and leaf (0.1mg/mL) extracts from avocado *P. americana* var. Hass induces apoptosis in Jurkat lymphoblastic leukemia cells, which are eliminated through an oxidative stress mechanism.

The levels of glutathione, potent antioxidant tripeptide, are found in higher concentrations in avocado when compared to matrices of other fruits (Flagg et al., 1994), and there is scientific evidence that an increase in the consumption of glutathione promotes the reduction of the risk of the development of larynx, pharynx and oral cavity cancers (Castillo-Juarez et al., 2009). A study performed by D'Ambrosio et al. (2011) identified and isolated from avocado mesocarp two aliphatic acetogenins. The chloroform extract of these two bioactive soluble compounds inhibited the growth of human oral cancer.

One novel alkene lactone was isolated from *Persea americana* Mill root bark and characterized (Falodun et al., 2013). The compound was subjected to anticancer activity using breast cancer cell line MCF-7 and the non-tumorigenic mammary epithelial cell line MCF-12 as a control for healthy tissue. At a concentration of 10 μg/mL, a significant reduction of proliferation of MCF-7 was induced while MCF-12 A cell was significantly stimulated by ten μg/mL. This opposite effect on both cell lines offers new possibilities in anticancer treatment because it would be excellent to have a drug that causes a reduction in the growth of cancer cells as well as stimulates the proliferation of healthy cells at the same time (Falodun et al., 2013). Mbaveng et al. (2018) also demonstrated the cytotoxic potential of *Persea americana* bark, which can be purified in order to isolate potential cytotoxic phytochemicals. P. americana has acceptable anti-proliferative activities in a panel of human cancer cell lines, and methanolic extracts could also be used to combat multi-drug-resistant cancer cell lines.

Antimicrobial peptides, responsible for the plants' defensins, show selective cytotoxicity toward cancer cells. Nevertheless, their mechanisms of action are not entirely understood. Guzmán-Rodríguez (2016) assessed the cytotoxicity of the antimicrobial peptides from avocado (*Persea americana* var. drymifolia) on the MCF-7 cancer cell line (a breast cancer cell line). This study was the first that suggested that avocado defensins are cytotoxic to cancer cells, namely to MCF-7 cells via the induction of mitochondrial apoptosis, which suggests that this defensin could be a potential molecule in the treatment of cancer. Flores-Alvarez et al., (2018) also studied the cytotoxicity defensin from avocado (PaDef) on cancer cells, in this case, K562 chronic myeloid leukemia cells. They have concluded that PaDef induces apoptosis on K562 cells, without affecting the mitochondrial membrane potential, neither the transmembranal potential nor the release of intracellular calcium. This process is related to the activation of caspase eight and involves the participation of TNF-α, which is a novel property for a plant defensin. These studies sustain that these avocado antimicrobial peptides are an attractive alternative to combat cancer cells.

A mixture of avocadene and avocadyne, two 17-carbon polyhydroxylated fatty alcohols present in seed and pulp of the avocado, designated by avocatin B, demonstrated to possess potent anticancer activity by selectively targeting and eliminating leukemia stem cells (Lee et al., 2015). Avocatin B has cytotoxic activity in human primary acute myeloid leukemia without affecting normal peripheral blood stem cells. Mechanistic investigations indicated that cytotoxicity relied on mitochondrial localization, as cells lacking functional mitochondria or carnitine palmitoyltransferase I, the enzyme that facilitates mitochondria lipid transport, were insensitive to avocatin B (Ahmed et al., 2018). Besides, avocatin B inhibited fatty acid oxidation and decreased NADPH levels, resulting in ROS-dependent leukemia cell death characterized by the release of mitochondrial proteins, apoptosis-inducing factor, and cytochrome c (Lee et al., 2015; Ahmed et al., 2018).

Beta-sitosterol (β-SITO), a phytosterol present in avocado, has been considered a potent anti-cancer agent that interferes with microtubule assembly dynamics by binding to a novel site on tubulin (Mahaddalkar et al., 2015). β-SITO is orally bioavailable and is considered to have no undesired side effects (Mahaddalkar et al., 2015). Extracts of avocado seeds also showed anticancer activity: methanol extracts have evidenced induction of apoptosis of MDA-MB-231 cells (Abubakar et al., 2017) and triterpenoid compounds isolated from ethanol extract have significant cytotoxic activity against breast and liver cells MCF-7 and HepG2, respectively (Ambrosio et al., 2011).

3.3. Antimicrobial Activity

The scientific community already knows the antifungal properties of avocado for about some decades. Avocado toxicity is due to compounds like diene (Z, Z)-2-hydroxy-4-oxohenicosa-12,15-dien-1-yl acetate (persin), that exist both in young and mature leaves (Carman and Handley, 1999) or 1,2,4-trihydroxyheptadec-16-yne, 1,2,4-trihydroxyheptadec-16-ene and 1-acetoxy-2,4-dihydroxyheptadec-16-yne isolated from the peel of immature fruit (Adikaram et al., 1992). Leite et al. (2009) tested the toxicity of extracts of avocado seeds against Artemia salina, evaluating their larvicidal activity related to their higher larvicidal activity towards Aedes aegypti, in the hexane extract, with the higher concentration of β-sitosterol and 1,2,4-trihydroxy-nonadecane. This group also showed antifungal activity of avocado seed extracts against Candida spp, Cryptococcus neoformans, and Malassezia pachydermatis, yeasts of great importance for human and veterinary medicine. On the other hand, flavonoids are known to be synthesized by plants in response to microbial infections exhibiting these compounds' antimicrobial activities (Melgar et al., 2017; Sansano et al., 2017). Melgar et al., (2018) studied the anti-bacterial and antifungal activity of *P. americana* by-products (peel and seed) against the following strains: *Staphylococcus aureus*, *Bacillus cereus*, *Micrococcus flavus* and *Listeria mono-cytogenes* (Gram-positive bacteria); *Escherichia coli*, *Pseudomonas aeruginosa*, *Salmonella typhimurium* and *Enterobacter cloacae* (Gram-negative bacteria); *Aspergillus fumigatus*, *Aspergillus ochraceus*, *Aspergillus versicolor*, *Aspergillus niger*, *Penicillium funiculosum*, *Penicillium ochrochloron*, *Penicillium verrucosum* var. *cyclopium* and *Trichoderma viride* (microfungi used for antifungal assays). *P. americana* by-products have revealed high anti-bacterial activity against all the gram-positive and gram-negative bacteria studied, displaying the extracts from seeds worst activity only against *E. cloacae* strain. Relativity to antifungal activity, these by-products extracts show moderate activity against the strains studied. The better fungicidal effect was against *Trichoderma viride*. Boadi et al. (2015) also studied the antimicrobial activities of methanol, ethyl acetate, chloroform, and petroleum ether leave extracts of *P. americana*. All the extracts were potent against the test microorganisms (*Bacillus subtilis*, *Escherichia coli*, *Salmonella typhi*, *Streptococcus aureus*, *Pseudomonas aeruginosa*, and *Candida albicans*), displaying the methanolic extract the most potent antimicrobial activities with the lowest minimum inhibitory concentration. This difference may be related to different degrees of solubility of the solvents for different phytoconstituents.

Avocado has also shown antibacterial activity against *Helicobacter pylori*, which potentiates gastric ulcers associated with the risk of gastric cancer (Liu et al., 2018). Methanol extract from avocado showed anti-*H. Pylori* activity at the lowest minimum inhibitory concentration among 53 Mexican traditional medicinal plants tested *in vitro* (Castillo-Juárez et al., 2009).

Rodríguez-Carpena et al. (2011) have studied antimicrobial properties of Hass and Fuerte varieties of avocado. The antimicrobial study with avocado pulp and by-products extracts from the two studied varieties show that Gram-positive bacteria were generally more sensitive than Gram-negative, being Gram-positive *B. cereus* and *L. monocytogenes* the most sensitive. Among tested Gram-negative, bacteria *E. coli* was the most sensitive. This antimicrobial potential of agricultural by-products could be a preliminary response for the substitution of antibiotics in animal-production farms. In past years there is an increasing trend in the use of antibiotics and other growth promoters in farm animals in tropical areas. Fortunately, as a substitute to antibiotics, these countries have considerable antimicrobial potential in plants by-products, source of a diverse pool of bioactive compounds with antimicrobial properties, which could be employed as feed supplements to improve animal health. Among others, high activity has been detected for avocado seeds (Guil-Guerrero et al., 2016).

3.4. Bodyweight Reduction

Avocado has various health benefits, including bodyweight reduction. Padmanabhan and Ar-umugam (2015) developed a study that proved that a hydro-alcoholic extract of avocado reduced mass index, adiposity index, total fat-pad mass, blood cholesterol, triglycerides, and LDL in rats fed with a high-fat diet. Their results also show that avocado extract has a profound influence on leptin activity, which controls satiety and hunger to regulate food intake. The same authors, in a previous study, investigated the effect of a hydro-alcoholic fruit extract of *P. americana* on the level of blood lipids, glutathione, lipid peroxidation products, adiponectin and peroxisome proliferator-activated receptor (PPAR)-γ expressions in rats fed a high-fat diet. Their results showed that body mass index, total fat pad mass, and adiposity index were significantly decreased in rats that have received the extract. This suggests that hydro-alcoholic avocado extract exhibits hypolipidemic activity prob-ably by increasing the mRNA expression of adiponectin and PPAR-γ, which reduces the risk of hyperlipidemia and obesity (Padmanabhan and Arumugam, 2014).

Othman et al. (2018) replaced butter by avocado purée and studied the influence of this change on nutritional, fatty acid, and organoleptic properties of low-fat muffins. They have prepared muffins with 25%, 50%, 75%, and 100% avocado purée and a control muffin, which was made with 100% butter. Muffins incorporated with avocado purée revealed a significant increase with respect carbohydrates; however, no significant changes were detected in all muffin formulations for protein and dietary fiber content, in comparison with the control sample. Fat content and caloric value of muffins incorporated with avocado purée were significantly decreased, showing an increment in the monounsaturated fatty acids content. Muffins were acceptable when up to 50% butter substitution, but a higher percentage led to undesirable flavor and aftertaste.

The potential use of avocado pulp as a fiber source has also been investigated in rats (Naveh et al., 2002). The total dietary fiber content of fresh avocado fruit of the Ettinger variety was 5.2 g/100 g. Food consumption and body weight gain were lower in rats fed avocado pulp compared with that fed cellulose. These data suggest that avocado pulp interferes with hepatic metabolism and causes an appetite depressant (Naveh et al., 2002).

In addition to dietary fiber, avocados contain monounsaturated fatty acids, essential nutrients, and phytochemicals. Based on this, a study was conducted with US adults of both genders, between 2001 and 2008, to investigate the relationships between avocado consumption and overall diet quality, energy and nutrient intakes, physiological indicators of health, and risk of metabolic syndrome. The results of this study proved that avocado consumers improved overall diet quality with significantly higher intakes of vegetables, fruit, total fat, monounsaturated and polyunsaturated fats, dietary fiber, vitamins E and K, magnesium, potassium; and lower intakes of added sugars. Bodyweight decreased significantly, and HDL increased, being the odds ratio for metabolic syndrome 50% lower in avocado consumers vs. non-consumers (Fulgoni et al., 2013).

3.5. Potential Neuroprotective Effect

Avocado fruit hydro-alcoholic extract contains flavonoids (rutin, quercetin, luteolin), phenolic compounds (gallic acid, ellagic acid, vanillic acid), and phytosterols (beta-sitosterol, stigmasterol) (Padmanabhan and Arumugam, 2015). This high content of phytochemicals, especially antioxidants, has a potential neuroprotective effect. Antioxidant compounds present in avocado are unique in the suppressing of radical generation, and thus may be promising as effective neuropreventive agents, preventing and curing various neurodegenerative diseases related with oxidative stress, such as Alzheimer's, Parkinson's and Huntington's diseases and amyotrophic lateral sclerosis (Ameer, 2016).

Avocado leaves and seeds are used in traditional medicine for the treatment and management of Alzheimer's disease, but information about the mechanism of actions is limited. Oboh et al., (2016) investigated the effect of these leaves and seeds aqueous extracts on the activity of the enzymes linked with Alzheimer's disease (acetylcholinesterase and butyrylcholinesterase) and their antioxidant potential *in vitro*. The extracts inhibited both enzymes activity, having the seed extract the highest inhibitory effect. Leaf extract has exhibited higher phenolic content and radical scavenging abilities, but lower Fe chelation ability compared with that of the seed. The phytochemical analysis revealed the presence of saponins and terpenoids in both extracts, and total alkaloid profile higher in the seed than in the leaf extract. These results propose that anticholinesterase and antioxidant activities of avocado leaf and seed could be related to their phytoconstituents.

Eser et al. (2011) studied the protective effects of avocado-soybean unsaponifiable (ASU) on the prefrontal cortex after global brain ischemia/reperfusion injury in rats. In the ischemia, rats that have received ASU malondialdehyde and tumor necrosis factor-alpha (TNFα) levels, as well as the number of apoptotic neurons, have decreased, and superoxide dismutase activities have increased significantly. These results suggest that ASU might have antioxidant and neuroprotective effects.

3.6. Osteoarthritis

Osteoarthritis is a disease characterized by progressive deterioration of cartilage and joint function associated with premature aging, especially in overweight or obese people (body mass index □ 30). This deterioration can be triggered by oxidative and inflammatory stress, which, in turn, can cause an imbalance between biosynthesis and degradation of the extracellular matrix of the joints, leading to loss of function (Dinubile, 2010).

The standard pharmacologic treatment includes agents for control of pain and inflammation (non-steroidal anti-inflammatory drugs, analgesics including opioids, intraarticular corticosteroids) and the group of the symptomatic slow-acting drugs for osteoarthritides such as glucosamine sulfate, chondroitin sulfate, diacerein, an unsaponifiable extract of soybean and avocado administered orally and intraarticular hyaluronic acid (Hermann et al., 2017).

One study reported that fruits rich in lutein and zeaxanthin are associated with decreased risk of cartilage defects (Wang et al., 2007). Also, the presence of unsaponifiable compounds in avocado and soybeans exert synergistic effects, promoting antioxidant activity against inflammatory processes (Dinubile, 2010).

Catunda et al. (2016) investigated the effectiveness of an avocado-soybean unsaponifiable extract in patients with arthralgia and osteoarthritis of the temporomandibular joint. A randomized, double-blind, placebo-controlled trial was carried out with fourteen women diagnosed with the referred diseases. The women were divided into two groups: ASU group and the placebo group. The group who received the ASU extract had a decrease in pain symptoms and an improvement in the quality of life. Furthermore, a significant decrease in the use of rescue medication was found in the ASU group compared to the placebo group. Another study in real-life situations was performed in Poland. Głuszko and Stasiek (2016) assessed the effects of oral ASU capsules on pain relief and functional ability in patients while they were receiving a routine treatment for knee osteoarthritis. The majority of patients adherent to the ASU treatment showed gradual alleviation of joint pain, improvement in functional ability, and a significant reduction in non-steroidal anti-inflammatory drug intake.

A study performed by Al-Afify et al., (2018) proved that oral administration of avocado-soybean unsaponifiables ameliorated the histopathological changes of the knee caused by induced osteoarthritis. Avocado-soybean unsaponifiables improved the ultrastructural changes and reduced oxidative stress and inflammation in the articular cartilage and subchondral bone of articular cartilages of rats with induced osteoarthritis.

Another study determined whether the addition of avocado-soybean unsaponifiable (Arthrocen) to the drinking water of osteoarthritis rats reduced direct joint pain. Osteoarthritis was induced in rats by intra-articular injection of sodium monoiodoacetate. The joint damage induced by injection was severe and consistent with end-stage osteoarthritis. Arthrocen consumption attenuated joint edema, inflammation, and pain (Goudarzi et al., 2018).

A previous study investigated *in vivo* the effects of the treatment with avocado-soybean unsaponifiables on the development of osteoarthritic on dogs. In animals treated with the ASU, the size of macroscopic lesions on the tibial plateaus decreased, as well as the severity of cartilage lesions on both tibial plateaus and femoral condyles and the cellular infiltration in the synovium. Treatment with avocado-soybean unsaponifiables also reduced loss of subchondral bone volume and calcified cartilage thickness compared with placebo. Immunohistochemical analysis of cartilage revealed that ASU significantly reduced the level of inducible nitric oxide synthase and matrix metalloproteinase 13 in cartilage (Boileau et al., 2009).

4. Avocado in Dermocosmetic

Plant extracts and oils have been utilized for a variety of purposes throughout history and are commonly used in cosmetics and topical pharmaceuticals products. In the last decades, there has been a growing trend towards the use of these extracts by the cosmetic industry. Among the reasons mentioned for this phenomenon are the decline of credibility in synthetic products and the preference for vegetal derivatives. Plant oils are nowadays recognized for their effects on both skin diseases and the restoration of cutaneous homeostasis. These oils also act as a protective barrier, allowing the skin to preserve moisture, with a decrease of transepidermal water loss. Furthermore, topical products have the benefit of higher bioavailability in the skin and having a localized effect rather than systemic effects (Kole et al., 2005; Lin et al., 2018).

Lipids are widely used by dermo-cosmetic, taking various roles, namely, bioactive agents, vehicles for the preparation of different formulations, nutritional supplements, and cosmetics, among others. Avocado oil has excellent applicability due to its emollient power and easy penetration into the skin, being absorbed more quickly than soy, corn, or almond (Athar and Nasir, 2005). Also, fatty acids increase the permeability of active substances in the skin and induce the diffusion properties of liposoluble preparations. Thus, the lipophilic extract of avocado may be a carrier of active substances in the formulation of creams, ointments, and other products for dermatological use (Stücker et al., 2001). The effect of *Persea americana* natural extracts on the subsequent dermal absorption and penetration of other cosmetics was evaluated by Muhammad et al., (2017). 14C-caffeine and 14C-salicylic acid were topically dosed with either 10% solutions of natural extracts or ethanol (control). The transdermal penetration of 14C-caffeine (hydrophilic) was significantly higher with avocado extract as compared to the control. Avocado extracts did not produce statistically significant effects on dermal penetration parameters for salicylic acid.

The avocado extract is also recommended as a basis for the preparation of various formulations that treat disorders and skin diseases, such as dermatitis or psoriasis. Stücker et al. (2001) performed a clinical trial to compare the effects of the vitamin D3 analog calcipotriol and vitamin B12 cream containing avocado oil, for use in the treatment of chronic plaque psoriasis. There was a more rapid development of beneficial effects with the use of calcipotriol in the initial eight weeks, but after twelve weeks, there were no significant differences between the two treatments. Although the efficacy of the calcipotriol preparation reached a maximum in the first four weeks and then began to subside, the benefits of the vitamin B12 cream with avocado oil remained at a constant level over the whole observation period, suggesting that this formulation is well-tolerated and may be suitable for use in long-term therapy.

Avocado oil is an excellent source of enrichment for dry, damaged, or chapped skin (Nayak, 2008). Another attractive property is its ability to absorb perfumes, great added value for cosmetic use. All these properties make it useful for cosmetics, being used in the preparation of moisturizing creams, lipsticks, sunscreens, exfoliants, and, due to the ease with which it forms emulsions, it becomes ideal for producing soaps (Tango et al., 2004). In addition to these properties, avocado has several bioactive compounds, namely, monounsaturated fatty acids, unsaponifiable fatty acids (phytosterols, tocopherols, squalene and furans lipids), vitamins A, C and E, carotenoids, phenolic agents, which enhances their use by dermo-cosmetics as cosmeticaceuticals (Nayak et al., 2008).

Avocado can be beneficial in combating aging of the skin based on the moisturizing and emollient properties of its oil and the considerable amount of unsaponifiable compounds, which act in the fight against senescence agents, favoring the skin's trophicity (Goetz, 2005).

Oliveira et al. (2013) studied the composition of avocado oil extracted from the pulp of the fruit and the effect of a semisolid formulation of that oil on wound healing in rats. Their results proved that avocado oil contains essential fatty acids (oleic (47.20%), palmitic (23.66%), linoleic (13.46%) docosadienoic (8.88%), palmitoleic (3.58%), linolenic (1.60%), eicosenoic (1.29%), and myristic acids (0.33%)). It also contains β-sitosterol, β-carotene, lecithin, minerals, and vitamins A, C, D, and E. A significant increase was observed in percentage wound contraction and reepithelialization in the groups treated with semisolid formulation or avocado oil compared to control. The topical application of avocado oil in rats also increased anti-inflammatory activity and collagen synthesis when compared to control groups.

Identical results were obtained by Nayak et al. (2008). Their study also aimed to evaluate the effect of topical and oral administration of avocado extract on excision and dead space wound models. The rats used in the excision wound model were divided into four groups of five each and received either topical or oral treatment. The animals used in the dead space wound model were divided into two groups of five each and were treated orally. Healing was assessed by the rate of wound contraction, period of epithelialization, granulation tissue weight, and hydroxyproline content. In the excision wound model, full epithelialization was observed on average on day 14 in the rats who receive oral or topical treatment.

In contrast, the controls took approximately 17 days to heal completely. The extract-treated wounds were found to epithelialize faster than the controls. Wet and dry granulation tissue weight and the hydroxyproline content of the tissue obtained from extract-treated animals used in the dead space wound model were significantly higher compared with the controls.

Oryan et al. (2015) investigated the effects of avocado-soybean unsaponifiables on the healing response of cutaneous wound defects in rats. Treatment increased tissue glycosaminoglycans and collagen contents at various stages of wound healing modulated inflammation, improved fibroplasia, and produced high amounts of scar tissue in the short term. In the long term, treatment reduced the scar tissue size, improved the quality, the rate of wound contraction, and the reepithelisation, being the treated lesions cosmetically more agreeable.

Skin exposure to ultraviolet radiation contributes to photoaging and the development of skin cancer by DNA mutations. *Persea americana* demonstrates a photoprotective effect against UVB radiation. Deuschle et al., (2018) investigated the antinociceptive and anti-inflammatory effects from a topical formulation containing the *P. americana* leaf extract in a UVB irradiation-induced burn model in mice and performed a gel-formulation stability study. The stability of the gel was studied, analyzing organoleptic characteristics, pH, and viscosity. The results proved that gel-formulation guaranteed stability, quality, and therapeutic effect, being able to prevent the UVB irradiation. This effect could be attributed to gel biological constituents, especially catechin, chlorogenic acid, and rutin. Polyhydroxylated fatty alcohols extracted from seed and pulp of avocados also have photo-protective activity against UV-induced damage in skin cells (Rosenblat et al., 2011). These compounds extracted from avocado suppress inflammatory response and provide non-sunscreen protection against UV-induced damage in skin cells. Introducing polyhydroxylated fatty alcohols to keratinocytes previous to exposure to UVB radiation exercised a protective effect, rising cell viability, decreasing the secretion of interleukin six and prostaglandin E2, and increasing DNA repair.

The peptides present in avocado also have some importance for the dermatology and dermo-cosmetic area, due to its antipruritic properties. Itching is known as an unpleasant sensation that causes the urge to itch. The origin of this phenomenon is in the dermal mast cells, and mediators of these cells, such as histamine or tryptase, stimulate the terminals of the C neurons, thus inducing the sensation of pruritus. Also, pruritus is known to be related to dry skin. Thus, avocado pep-tides can attenuate this sensation by several processes: because of their ability to inhibit the release of histamine and tryptase by the mast cells; decrease the genetic expression at the epidermal level of the receptor involved in the release of these mediators from pruritus; positively affect the ceramides in the epidermis and the synthesis of fatty acids, essential for epidermal balance; and, ultimately, significantly increase the production of hyaluronic acid and glycosaminoglycans by the keranocytes, which are essential for the various layers of the skin and its hydrolipidic film (Bredif et al., 2010).

Conclusion

In the context of a healthy diet, consumption of avocados can fit into a full range of healthy eating plans. The primary purpose of this review is to promote the production and processing of this fruit and its by-products (peel and seed), highlighting the possibility of extracting bioactive compounds and the possibility to use them as natural additives for the food industry as well as pharmaceutical and cosmetics industries. These benefits will open up as a scope for future utilization of fruit and vegetable waste for therapeutic and nutraceuticals purposes.

The edible part of avocado (pulp) has an essential quantity of hydroxybenzoic and hydroxycinnamic acids and procyanidins. However, the peel and the seed of avocado varieties present higher amounts of phenolic compounds, namely catechins, procyanidins, and hydroxycinnamic acids, displaying more antioxidant potential. In this work, it is also suggested that the reuse of avocado seed and peel can be powerful sources of natural ingredients and additives to provide new technological solutions through the application of green technology. The pulp of the avocado fruits, as well as their by-products, are rich sources of lipids, starch and fibers, and polyphenolics proving to be a potential natural source of bioactive ingredients for use in the food, cosmetic and pharmaceutical industries.

References

Abubakar, A. N. F., Achmadi, S. S., and Suparto, I. H. (2017). Triterpenoid of avocado (*Persea americana*) seed and its cytotoxic activity toward breast MCF-7 and liver HepG2 cancer cells. *Asian Pacific Journal of Tropical Biomedicine*, 7(5): 397-400.

Adeboye, J. O., Fajonyomi, M. O., Makinde, J. M., and Taiwo, O. B. (1999). A preliminary study on the hypotensive activity of *Persea americana* leaf extracts in anesthetized, normotensive rats. *Fitoterapia,* 70(1): 15-20.

Adeyemi, O. O., Okpo, S. O., and Ogunti, O. O. (2002). Analgesic and anti-inflammatory effects of the aqueous extract of leaves of *Persea americana* Mill (Lauraceae). *Fitoterapia*, 73(5): 375-380.

Adikaram, N. K. B., Ewing, D. F., Karunaratne, A. M., and Wijeratne, E. M. K. (1992). Antifungal compounds from immature avocado fruit peel. *Phytochemistry,* 31(1): 93-96.

Ahmed, N., Smith, R. W., Henao, J. J. A., Stark, K. D., and Spagnuolo, P. A. (2018). Analytical Method to Detect and Quantify Avocatin B in Hass Avocado Seed and Pulp Matter. *Journal of Natural Products,* 81(4): 818-824.

Al-Afify, A. S. A., El-Akabawy, G., El-Sherif, N. M., El-Safty, F. E. A., and El-Habiby, M. M. (2018). Avocado soybean unsaponifiables ameliorates cartilage and subchondral bone degeneration in mono-iodoacetate-induced knee osteoarthritis in rats. *Tissue and Cell*, 52: 108-115.

Almeida, A. P., Miranda, M. M. F. S., Simoni, I. C., Wigg, M. D., Lagrota, M. H. C., and Costa, S. S. (1998). Flavonol monoglycosides isolated from the antiviral fractions of *Persea americana* (Lauraceae) leaf infusion. *Phytotherapy Research*, 12(8): 562-567.

Ambrosio, S. M. D., Han, C., Pan, L., Kinghorn, A. D., and Ding, H. (2011). Aliphatic acetogenin constituents of avocado fruits inhibit human oral cancer cell proliferation by targeting the EGFR/RAS/RAF/MEK/ERK1/2 pathway. *Biochemical and Biophysical Research Communications,* 409(3): 465-469.

Ameer, K. (2016). Avocado as a major dietary source of antioxidants and its preventive role in neurodegenerative diseases. *Advances in Neurobiology*,12: 337-354.

Anaka, O. N., Ozolua, R. I. and Okpo, S. O. (2009). Effect of the aqueous seed extract of *Persea americana* Mill (Lauraceae) on the blood pressure of spraguedawley rats. *African Journal of Pharmacy and Pharmacology,* 3(10): 485-490.

Antia, B. S., Okokon, J. E., and Okon, P. A. (2005). Hypoglycemic activity of aqueous leaf extract of *Persea americana* Mill. *Indian Journal of Pharmacology,* 37(5): 325-326.

Araújo, R. G., Rodriguez-Jasso, R. M., Ruiz, H. A., Pintado, M. M. E., and Aguilar, C. N. (2018). Avocado by-products: Nutritional and functional properties. *Trends in Food Science & Technology*, 80: 51-60.

Athar, M., and Nasir, S. M. (2005). Taxonomic perspective of plant species yielding vegetable oils used in cosmetics and skin care products. *African Journal of Biotechnology*, 4(1): 36-44.

Awad, A. B., and Fink, C. S. (2000). Phytosterols as anticancer dietary components: evidence and mechanism of action. *The Journal of Nutrition,* 130(9): 2127–2130.

Ballmann, C., and Müeller, B. W. (2008). Stabilizing effect of cetostearyl alcohol and glycerylmonstearate as co-emulsifiers on hydrocarbon-free O/W glyceride creams. *Pharmaceutical Development and Technology*, 13(5): 433-445.

Barbosa-Martin, E., Chel-Guerrero, L., Gonzalez-Mondragon, E., and Betancur-Ancona, D. (2016). Chemical and technological properties of avocado (*Persea americana* Mill.) seed fibrous residues. *Food and Bioproducts Processing*, 100: 457-463.

Bes-Rastrollo, M., van Dam, R. M., Martinez-Gonzalez, M. A., Li, T. Y., Sampson, L. L., and Hu, F. B. (2008). Prospective study of dietary energy density and weight gain in women. *American Journal of Clinical Nutrition*, 88(3): 769-767.

Bill, M., Sivakumar, D., Thompson, A. K., and Korsten, L. (2014). Avocado fruit quality management during the postharvest supply chain. *Food Reviews International*, 30(3): 169-202.

Boadi, N. O., Saah, S. A., Mensah, J. K., Badu, M., and Addai-Arhinand, S. (2015). Phytoconstituents, antimicrobial and antioxidant properties of the leaves of *Persea americana* Mill cultivated in Ghana. *Journal of Medicinal Plants Research*, 9(36): 933-939.

Boileau, C., Martel-Pelletier, J., Caron, J., Msika, P., Guillou, G. B., Baudouin, C., and Pelletier, J. P. (2009). Protective effects of total fraction of avocado-soybean unsaponifiables on the structural changes in experimental dog osteoarthritis: inhibition of nitric oxide synthase and matrix metalloproteinase-13. *Arthritis Research & Therapy*,11(2): R41.

Bonilla-Porras, A., Salazar-Ospina, A., Jimenez-Del-Rio, M., Pereaněz-Jimenez, A., and Velez-Pardo, C. (2014). Pro-apoptotic effect of *Persea americana* var. Hass (avocado) on Jurkat lymphoblastic leukemia cells. *Pharmaceutical Biology*, 2014; 52(4): 458-465.

Brai, B., Adisa, R. A., and Odetola, A. A. (2014). Hepatoprotective properties of aqueous leaf extract of *Persea americana*, Mill (Lauraceae) 'avocado' against CCL4-induced damage in rats. African *Journal of Traditional, Complementary and Alternative Medicines*,11(2): 237-244.

Bredif, S., Baudouin, C., and Msika, P. (2010). Antiitching properties of patented avocado peptides. *Journal of the American Academy of Dermatology*, 62(3): Supplement 1, AB54.

Calderón-Oliver, M., Escalona-Buendía, H. B., Medina-Campos, O. N., Pedraza-Chaverri, J., Pedroza-Islas, R., and Ponce-Alquicira, E. (2016). Optimization of the antioxidant and antimicrobial response of the combined effect of nisin and avocado by-products. *Food Science* and Technology, 65: 46-52.

Carman, R. M., and Handley, P. N. (1999). Antifungal diene in leaves of various avocado cultivars.*Phytochemistry*, 50(8): 1329-1331.

Carvajal-Zarrabal, O., Nolasco-Hipolito, C., Aguilar-Uscanga, M. G., Melo-Santiesteban, G., Hayward-Jones, P. M. and Barradas-Dermitz, D. M. (2014). Avocado oil supplementation modifies cardiovascular risk profile markers in a rat model of sucrose-induced metabolic changes. Dis Markers, Article ID 386425.

Castillo-Juárez, I., González, V., Jaime-Aguilar, H., Martínez, G., Linares, E., Bye, R. and Romero, I. (2009). Anti-Helicobacter pylori activity of plants used in Mexican traditional medicine for gastrointestinal disorders. *Journal of Ethnopharmacology*, 122(2): 402-405.

Catunda, I. S., Vasconcelos, B. C. E., Andrade, E. S. S., and Costa, D. F. N. (2016). Clinical effects of an avocado–soybean unsaponifiable extract on arthralgia and osteoarthritis of the temporomandibular joint: preliminary study. *International Journal of Oral and Maxillofacial Surgery*, 45(8): 1015-1022.

Chen, H., Morrell, P. L., Ashworth, V. E. T. M., de la Cruz, M., and Clegg, M. T. (2009). Tracing the geographic origins of major avocado cultivars. *Journal of Heredity,* 100: 56-65.

Christian, E. O., Okwesili, F. N., Parker, E. J., and Okechukwu, P. U. (2014). Acute toxicity investigation and anti-diarrhoeal effect of the chloroform-methanol extract of the leaves of *Persea americana*. *Iranian Journal of Pharmaceutical Research*, 13(2): 651-658.

Chrysohoou, C., and Stefanadis, C. (2013). Longevity and Diet. Myth or pragmatism? *Maturitas,* 76(4): 303-307.

Contreras-Gutiérrez, P. K., Hurtado-Fernández, E., Gómez-Romero, M., Hormaza, J. I., Carrasco-Pancorbo, A. and Fernández-Gutiérrez, A. (2013). Determination of changes in the metabolic profile of avocado fruits (*Persea americana*) by two CE-MS approaches (targeted and non-targeted). *Electrophoresis,* 34: 2928-2942.

Corzzini, S. C. S., Barros, H. D. F. Q., Grimaldi, R., and Cabral, F. A. (2017). Extraction of edible avocado oil using supercritical CO2 and a CO2/ethanol mixture as solvents. *Journal of Food Engineering*, 194: 40-45.

Cowan A. K. and Wolstenholme, B. N. (2003). Avocados. *Encyclopedia of Food Sciences and Nutrition,* Elsevier, Amsterdam.

Cummings, K., and Schroeder, C. A. (1942). *Anatomy of the Avocado Fruit*. California Avocado Society, 1942 Yearbook 27: 56-64.

D'Ambrosio, S. M., Han, C., Pan, L., Kinghorn, A. D., and Ding, H. (2011). Aliphatic acetogenin constituents of avocado fruits inhibit human oral cancer cell proliferation by targeting the EGFR/RAS/RAF/MEK/ERK1/2 pathway. *Biochemical and Biophysical Research Communications*, 409(3): 465–469.

Dabas, D., Shegog, R. M., Ziegler, G. R., and Lambert, J. D. (2013). Avocado (*Persea americana*) seed as a source of bioactive phytochemicals. *Current Pharmaceutical Design*, 19(34): 6133-6140.

Dávila, J. A., Rosenberg, M., Castro, E., and Cardona, C. A. (2017). A model biorefinery for avocado (*Persea americana* Mill.) processing. *Bioresource Technology*, 243: 17-29.

De Bon, H. (2010). Fruitrop, Nº. 183, CIRAD Market News Service: Montpellier, France.

Dembitsky, V. M., Poovarodom, S., Leontowicz, H., Leontowicz, M., Vearasilp, S., Trakhtenberg, S., and Gorinstein, S. (2011). The multiple nutrition properties of some exotic fruits: Biological activity and active metabolites. *Food Research International*, 44(7): 1671-1701.

Deuschle, V. C. K. N., Brusco, I., Piana, M., Faccin, H., de Carvalho, L. M., Oliveira, S. M., and Viana, C. (2018). *Persea americana* Mill.

crude extract exhibits antinociceptive effect on UVB radiation-induced skin injury in mice. *Inflammopharmacology,* 27(2): 1-16.

Ding, H., Han, C., Guo, D., Chin, Y. W., Ding, Y., Kinghorn, A. D., and D'Ambrosio S. M. (2009). Selective induction of apoptosis of human oral cancer cell lines by avocado extracts via a ROS-mediated mechanism. *Nutrition and Cancer*, 61(3): 348-56.

DiNubile, N. A. (2010). A potential role for avocado- and soybean-based nutritional supplements in the management of osteoarthritis: a review. *The Physician and Sportsmedicine*, 38(2): 71-81.

Dreher, M. L., and Davenport, A. J. (2013). Hass avocado composition and potential health effects. *Critical Reviews in Food Science and Nutrition,* 53: 738-750.

Duarte, P. F., Chaves, M. A., Borges, C, D., and Mendonça, C. R. B. (2016). Avocado: characteristics, health benefits, and uses. *Food Technology*, 46(4): 747-754.

Duresa, L. W. (2017). Phytochemical screening and antioxidant activity of selected mango (Mangifera indica L.) and avocado (*Persea americana*) fruits in Illu Ababor zone, Oromia regional state, Ethiopia. *Journal of Applied Chemistry*, 10(5): 24-28.

Eggersdorfer, M., and Wyss, A. (2018). Carotenoids in human nutrition and health. *Archives of Biochemistry and Biophysics*, 652: 18-26.

Eser, O., Songur, A., Yaman, M., Cosar, M., Fidan, H., and Sahin, O. (2011). The protective effect of avocado soybean unsaponifilables on brain ischemia/reperfusion injury in rat prefrontal cortex. *British Journal of Neurosurgery*, 25(6): 701-706.

Falodun, A., Engel, N., Kragl, U., Nebe, B., and Langer, P. (2013). Novel anticancer alkene lactone from *Persea americana*. *Pharmaceutical Biology,* 51(6): 700-706.

Figueroa, J. G., Borrás-Linares, I., Lozano-Sánchez, J. and Segura-Carretero, A. (2018a). Comprehensive characterization of phenolic and other polar compounds in the seed and seed coat of avocado by HPLC-DAD-ESI-QTOF-MS. *Food Research International*, 105: 752-763.

Figueroa, J. G., Borrás-Linares, I., Lozano-Sánchez, J., and Segura-Carretero, A. (2018b). Comprehensive identification of bioactive compounds of avocado peel by liquid chromatography coupled to ultra-high-definition accurate-mass QTOF. *Food Chemistry*, 245: 707-716.

Flagg, E. W., Coates, R. J., Jones, D. P., Byers, T. E., Greenberg, R. S., Gridley, G., McLaughlin, J. K., Blot, W. J., Haber, M., Preston-Martin, S., et al., (1994). Dietary glutathione intake and the risk of oral and pharyngeal cancer. *American Journal of Epidemiology*, 139(5): 453-65.

Flores-Alvarez, L. J, Guzmán-Rodríguez, J. J., López-Gómez, R., Salgado-Garciglia, R., Ochoa-Zarzosa, A. and López-Meza, J. E. (2018). PaDef defensin from avocado (*Persea americana* var. drymifolia) is cytotoxic to K562 chronic myeloid leukemia cells through extrinsic apoptosis. *International Journal of Biochemistry and Cell Biology*, 99: 10-18.

Food and Agriculture Organization (FAO), United Nations. *The State of Food Security in the World: Addressing Food Insecurity in Protracted Crisis*; FAO: Rome, 2010.

Forero-Doria, O., García, M. F., Vergara, C. E., and Guzman, L. (2017). Thermal analysis and antioxidant activity of oil extracted from pulp of ripe avocados. *Journal of Thermal Analysis and Calorimetry*, 130: 959-966.

Fulgoni, V. L., Dreher, M., and Davenport, A. J. (2013). Avocado consumption is associated with better diet quality and nutrient intake, and lower metabolic syndrome risk in US adults: results from the National Health and Nutrition Examination Survey (NHANES) 2001-2008. *Nutrition Journal*, 12: 1-6.

Furlan, C. P. B., Valle, S. C., Östman, E., Maróstica Júnior, M. R., Tovar J. (2017). Inclusion of Hass avocado-oil improves postprandial metabolic responses to a hypercaloric-hyperlipidic meal in overweight subjects. *Journal of Functional Foods*, 38: 349-354.

Ghasemzadeh, A., and Ghasemzadeh, N. (2011). Flavonoids and phenolic acids: Role and biochemical activity in plants and humans. *Journal of Medicinal Plant Research*, 5: 6697-6703.

Głuszko, P., and Stasiek, M. (2016). Symptom-modifying effects of oral avocado-soybean unsaponifiables in routine treatment of knee osteoarthritis in Poland. An open, prospective observational study of patients adherent to a 6-month treatment. *Reumatologia,* 54(5): 217-226.

Goetz, P. (2005). Élements du traitement par phytothérapie du vieillissement de la peau et du conjoctif. *Phytothérapie Clinique*, 3(2): 72-76.

Golukcu, M., and Ozdemir. F. (2010). Changes in phenolic composition of avocado cultivars during harvesting time. Chemistry of Natural *Compounds,* 46(1): 112-115.

Gómez, F. S., Sánchez, S. P., Iradi, M. G. G., Azman, N. A. M. and Almajano, M. P. (2014). Avocado seeds: extraction optimization and possible use as antioxidant in food. *Nutrients*, 3(2): 439-454.

Goudarzi, R., Reid, A., and McDougall, J. J. (2018). Evaluation of the novel avocado-soybean unsaponifiable Arthrocen to alter joint pain and inflammation in a rat model of osteoarthritis. *PLoS ONE*, 13(2): e0191906.

Guil-Guerrero, J. L., Ramos, L., Moreno, C., Zúñiga-Paredes, J. C., Carlosama-Yepez, M. and Ruales, P. (2016). Antimicrobial activity of plant-food by-products: A review focusing on the tropics. *Livestock Science*, 189: 32-49.

Guzmán-Rodríguez, J. J., López-Gómez, R., Salgado-Garciglia, R., Ochoa-Zarzosa, A. and López-Meza, J. E. (2016). The defensin from avocado (*Persea americana* var. drymifolia) PaDef induces apoptosis in the human breast cancer cell line MCF-7. *Biomedicine & Pharmacotherapy,* 82: 620-627.

Gyles, C. L., Carlberg, J. G., Gustafson, J., Davlut, D. A., and Jones, P. J. H. (2010). Economic valuation of the potential health benefits from foods enriched with plant sterols in Canada. *Food and Nutrition Research,* 54: 1-6.

Haque, M., and Mahadeva, R. (2012). Screening of secondary metabolites and comparative superoxide radical scavenging along with reducing potency of *Persea americana* using various solvent extracts. *Drug Invention Today*, 4(11): 569-574.

Hermann. W., Lambova, S., and Müller-Ladner, U. (2017). Current treatment options for osteoarthritis. *Current Rheumatology Reviews*, 14(2): 108-116.

Hurtado-Fernández, E., Contreras-Gutiérrez, P., Cuadros-Rodríguez, L., Carrasco-Pancorbo, A. and Fernández-Gutiérrez, A. (2013). Merging a sensitive capillary electrophoresis-ultraviolet detection method with chemometric exploratory data analysis for the determination of phenolic acids and subsequent characterization of avocado fruit. *Food Chemistry,* 141: 3492–3503.

Hurtado-Fernández, E., Pacchiarotta, T., Mayboroda, O. A., Fernández-Gutiérrez, A. and Carrasco-Pancorbo A. (2014). Quantitative characterization of important metabolites of avocado fruit by gas chromatography coupled to different detectors (APCI-TOF MS and FID). *Food Research International*, 62: 801-811.

Hurtado-Fernández, E., Pacchiarotta, T., Mayboroda, O. A., Fernández-Gutiérrez, A. and Carrasco-Pancorbo, A. (2015). Metabolomic analysis of avocado fruits by GC-APCI-TOF MS: Effects of ripening degrees and fruit varieties. *Analytical and Bioanalytical Chemistry*, 407: 547-555.

Jimenez-Garcia, S. N., Vazquez-Cruz, M. A., Garcia-Mier, L., Contreras-Medina, L. M., Guevara-González, R. G., Garcia-Trejo, J. F. and Feregrino-Perez, A. A. (2018). Chapter 13 - Phytochemical and Pharmacological Properties of Secondary Metabolites in Berries. *Therapeutic Foods Handbook of Food Bioengineering*, 397-427.

Kaur, H. K., Mehta, R., Tandey, R., Sen, K. K., and Mandal, V. (2016). Ten years of research on phenolics (2005–2015): A status report. Pacific Science Review A: *Natural Science and Engineering*, 18(1): 1-4.

Knight, J. R. (2002). "History, Distribution, and Uses." In: Whiley, A. W., Schaffer, B., Wolstenholme, B. N. (Eds.). *Avocado: Botany, Production and Uses,* Cabi International, Wallingford.

Kole, P. L., Jadhav, H. R., Thakurdesai, P., and Nagappa, A. N. (2005). Cosmetics potential of herbal extracts. *Indian Journal of Natural Products and Resources*, 4(4): 315-321.

Kopec, R., and Failla, M. (2018). Recent advances in the bioaccessibility and bioavailability of carotenoids and effects of other dietary lipophiles. *Journal of Food Composition and Analysis*, 68: 16-30.

Kosińska, A., Karamać, M., Estrella, I., Hernández, T., Bartolomé, B., and Dykes, G. A. (2012). Phenolic compound profiles and antioxidant capacity of *Persea americana* Mill. peels and seeds of two varieties. *Journal of Agricultural and Food Chemistry*, 60(18): 4613-4619.

Kritchevsky, D., and Chen, S. C. (2005). Phytosterols—health benefits and potential concerns: a review. *Nutrition Research*, 25(5): 413–428.

Krumreich, F. D., Borges, C., Mendonça, C., Jansen-Alvesa, C., and Zambiazi, R. (2018). Bioactive compounds and quality parameters of avocado oil obtained by different processes. *Food Chemistry*, 257: 376-381.

Kumar, A., Kumarchandra, R., Rai, R., and Sanjeev, G. (2017). Anticlastogenic, radiation antagonistic, and anti-inflammatory activities of *Persea americana* in albino Wistar rat model. *Research in Pharmaceutical Sciences*, 12(6): 488-499.

Larijani, L. V., Ghasemi, M., Kenari, S. A., and Naghshvar, F. (2014). Evaluating the effect of four extracts of avocado fruit on esophageal squamous carcinoma and colon adenocarcinoma cell lines in comparison with peripheral blood mononuclear cells. *Acta Medica Iranica,* 52(3): 201-205.

Lee, E. A., Angka, L., Rota, S. G., Hanlon, T., Mitchell, A., Hurren, R., et al.,(2015). Targeting mitochondria with avocatin B induces selective leukemia cell death. *Cancer Research*, 75(12): 2478-2488.

Leite, J. J. G., Brito, É. H. S., Cordeiro, R. A., Brilhante, R. S. N., Sidrim, J. J. C., Bertini, L. M., et al., (2009). Chemical composition, toxicity and larvicidal and antifungal activities of *Persea americana* (avocado) seed extracts. *Revista da Sociedade Brasileira de Medicina Tropical*, 42(2): 110-113.

Lima, C. R., Vasconcelos, C. F., Costa-Silva, J. H., Maranhao, C. A., Costa, J., Batista, T. M., Carneiro, E. M., Soares, L. A. L., Ferreira F. and Wanderley, A. G. (2012). Anti-diabetic activity of extract from *Persea americana* Mill leaf via the activation of protein kinase B

(PKB/Akt) in streptozotocin-induced diabetic rats. *Journal of ethnopharmacology,* 141(1): 517-525.

Lin, T. K., Zhong, L., and Santiago, J. L. (2018). Anti-Inflammatory and skin barrier repair effects of topical application of some plant oils. *International Journal of Molecular Sciences*, 19(1): pii: E70.

Liu, Q., Meng, X., Li, Y., Zhao, C.-N., Tang, G.-Y., Li, S., Gan, R.-Y. and Li, H.-B. (2018). Natural products for the prevention and management of Helicobacter pylori infection. *Comprehensive Reviews in FoodScience and Food Safety*, 17(4): 937-952.

López-Cobo, A., Gómez-Caravana, A. M., Pasini, F., Caboni, M. F., Segura-Carretero, A. and Férnandez-Gutiérrez, A. (2016). HPLC-DAD-ESI-QTOF-MS and HPLC-FLD-MS as valuable tools for the determination of phenolic and other polar compounds in the edible part and by-products of avocado. *LWT - Food Science and Technology*, 73: 505-513.

Lu, Q. Y., Zhang, Y., Wang, Y., Wang, D., Lee, R. P., Gao, K., Byrns, R., and Heber, D. (2009). California Hass avocado: profiling of carotenoids, tocopherol, fatty acid, and fat content during maturation and from different growing areas. *Journal of Agricultural and food chemistry*, 57(21): 10408-10413.

Mahaddalkar, T., Suri, C., Naik, P. K., and Lopus, M. (2015). Biochemical characterization and molecular dynamic simulation of β-sitosterol as a tubulin-binding anticancer agent. *European Journal of Pharmacology*, 760: 154-162.

Mahmassani, H. A., Avendano, E. E., Raman, G., and Johnson, E. J. (2018). Avocado consumption and risk factors for heart disease: a systematic review and meta-analysis. *The American Journal of Clinical Nutrition,* 107(4): 523-536.

Marlett, J. S., and Cheung, T.F. (1997). Database and quick methods of assessing typical dietary fiber intakes using data for 228 commonly consumed foods. *Journal of the American Dental Association*, 97: 1139-1148.

Márquez-Ramírez, C. A., de la Paz, J. L. H., Ortiz-Avila, O., Raya-Farias, A., González-Hernández, J. C., Rodríguez-Orozco, A. R., Salgado-

Garciglia, R., Saavedra-Molina, A., Godínez-Hernández, D. and Cortés-Rojo, C. (2018). Comparative effects of avocado oil and losartan on blood pressure, renal vascular function, and mitochondrial oxidative stress in hypertensive rats. *Nutrition,* 54: 60-67.

Mbaveng, A. T., Manekeng, H. T., Nguenang, G. S., Dzotam, J. K., Kuete, V., and Efferth, T. (2018). Cytotoxicity of 18 Cameroonian medicinal plants against drug sensitive and multi-factorial drug resistant cancer cells. *Journal of Ethnopharmacology*, 222: 21-33.

Melgar, B., Dias, M. I., Ciric, A., Sokovic, M., Garcia-Castello, E. M., Rodriguez-Lopez, A. D., Barros, L., and Ferreira, I. (2017). By-product recovery of Opuntia spp. peels: betalainic and phenolic profiles and bioactive properties. *Industrial Crops and Products*, 107, 353-359.

Melgar, B., Dias, M. I., Ciric, A., Sokovic, M., Garcia-Castello, E. M., Rodriguez-Lopez, A. D., Barros, L., and Ferreira, I. (2018). Bioactive characterization of *Persea americana* Mill. By-products: A rich source of inherent antioxidants. *Industrial Crops and Products*, 111: 212-218.

Meyer, M. D., and Terry, L. A. (2008). Development of a rapid method for the sequential extraction and subsequent quantification of fatty acids and sugar from avocado mesocarp tissue. *Journal of Agricultural and Food Chemistry*, 56:7439-7445.

Min, K., and Ebeler, S. E. (2008). Flavonoid effects on DNA oxidation at low concentrations relevant to physiological levels. *Food and Chemical Toxicology*, 46(1): 96-104.

Muhammad, F., Wiley, J., and Riviere, J. E. (2017). Influence of some plant extracts on the transdermal absorption and penetration of marker penetrants. *Cutaneous and Ocular Toxicology*, 36(1): 60-66.

Nagarajan, J., Ramanan, R. N., Raghunandan, M. E., Galanakis, C. M., and Krishnamurthy, N. P. (2017). Chapter 8 – Carotenoids. Nutraceutical and Functional Food Components. *Effects of Innovative Processing Techniques*, 259-296.

Naveh, E., Werman, M. J., Sabo, E., and Neeman, I. (2002). Defatted avocado pulp reduces body weight and total hepatic fat but increases plasma cholesterol in male rats fed diets with cholesterol. *Journal of Nutrition*, 132(7): 2015-2018.

Nayak, B. S., Raju, S. S., and Rao, A. V. C. (2008). Wound healing activity of *Persea americana* (avocado) fruit: a preclinical study on rats. *Journal of Wound Care*, 17(3): 123-126.

Nour, V., Stampar, F., Veberic, R., and Jakopic, J. (2013). Anthocyanins profile, total phenolics, and antioxidant activity of black currant ethanolic extracts as influenced by genotype and ethanol concentration. *Food Chemistry,* 141: 961-966.

Oboh, G., Odubanjo, V. O., Bello, F., Ademosun, A. O., Oyeleye, S. I., Nwanna, E. E., and Ademiluyi, A. O. (2016). Aqueous extracts of avocado pear (*Persea americana* Mill.) leaves and seeds exhibit anti-cholinesterases and antioxidant activities *in vitro*. *Journal of Basic and Clinical Physiology and Pharmacology*, 27(2): 131-140.

Oliveira, A. P., Franco, E. S., Barreto, R. R., Cordeiro, D. P., Melo, R. G., Aquino, C. M. F., et al., (2013). Effect of semisolid formulation of *Persea americana* Mill (Avocado) oil on wound healing in rats. *Evidence-Based Complementary and Alternative Medicine*, Article ID 472382.

Ortiz-Avila, O., Sámano-García, C. A., Calderón-Cortés, E., Pérez-Hernández, I. H., Mejía-Zepeda, R., Rodríguez-Orozco, A. R., Saavedra-Molina, A. and Cortés-Rojo, C. (2013). Dietary avocado oil supplementation attenuates the alterations induced by type I diabetes and oxidative stress in electron transfer at the complex II-complex III segment of the electron transport chain in rat kidney mitochondria. *Journal of Bioenergetics and Biomembranes*, 45: 271-287.

Oryan, A., Mohammadalipour, A., Moshiri, A., and Tabandeh, M. R. (2015). Avocado/soybean unsaponifiables: a novel regulator of cutaneous wound healing, modeling, and remodeling. *InternationalWound Journal*, 12(6): 674-685.

Othman, N. A., Abdul, M. M., Harith, S., and Wan, I. W. R. (2018). Influence of avocado purée as a fat replacer on nutritional, fatty acid, and organoleptic properties of low-fat muffins. *The Journal of the American College of Nutrition*, 37(7): 583-588.

Owolabi, M. A., Jaja, S. I., and Coker, H. A. (2005). Vasorelaxant action of aqueous extract of the leaves of *Persea americana* on isolated thoracic rat aorta. *Fitoterapia,* 76(6): 567-573.

Padmanabhan, M., and Arumugam, G. (2014). Effect of *Persea americana* (avocado) fruit extract on the level of expression of adiponectin and PPAR-γ in rats subjected to experimental hyperlipidemia and obesity. *Journal of Complementary and Integrative Medicine*, 11(2): 107-119.

Padmanabhan, M., and Arumugam, G. (2015). The modulating effect of *Persea americana* fruit extract on the level of expression of fatty acid synthase complex, lipoprotein lipase, fibroblast growth factor-21, and leptin – A biochemical study in rats subjected to experimental hyperlipidemia and obesity. *Phytomedicine,* 22(10): 939-945.

Pahua-Ramos, M. E., Garduño-Siciliano, L., Dorantes-Alvarez, L. Chamorro-Cevallos, G., Herrera-Martínez, J., Osorio-Esquivel, O. and Ortiz-Moreno, A. (2014). Reduced-calorie avocado paste attenuates metabolic factors associated with a hypercholesterolemic-high fructose diet in rats. *Plant Foods for Human Nutrition*, 69:18-24.

Pahua-Ramos, M. E., Ortiz-Moreno, A., Chamorro-Cevallos, G., Hernández-Navarro, M. D., Necoechea-Mondragón, H. and Hernández-Ortega, M. (2012). Hypolipidemic effect of avocado (*Persea americana* Mill) seed in a hypercholesterolemic mouse model. *Plant Foods for Human Nutrition*, 67(1): 10-16.

Paturi, G., Butts, C. A., and Bentley-Hewitt, K. L. (2017). Influence of dietary avocado on gut health in rats. *Plant Foods for Human Nutrition*, 72: 321-323.

Peou, S., Milliard-Hasting, B., and Shah, S. A. (2016). Impact of avocado-enriched diets on plasma lipoproteins: A meta-analysis. *Journal of Clinical Lipidology*, 10(1): 161–17.

Pérez-Méndez, O. and Hernández, G. L. (2007). High density lipoproteins (HDL) size and composition are modified in the rat by diet supplementation with "Hass" avocado. *Archivos de Cardíologia de Mexico,* 77(1): 17-24.

Piironen, V., Lindsay, D. G., Miettinen, T. A., Toivo, J., and Lampi, A. M. (2000). Plant sterols: biosynthesis, biological function, and their importance to human nutrition. *Journal of the Science of Food and agriculture*, 80(7): 939-966.

Prasad, S., Gupta, S. C., and Tyagi, A. K. (2017). Reactive oxygen species (ROS) and cancer: Role of antioxidative nutraceuticals. *Cancer Letters,* 387: 95-105.

Raharjo, S. H. T., Witjaksono, N. F. N., Gomez-Lim, M. A., Padilla G., Litz, R. E. (2008). Recovery of avocado (*Persea americana* Mill.) plants transformed with the antifungal plant defensin gene PDF1.2. *InVitro Cellular & Developmental Biology – Plant,* 44(4): 254-262.

Rajkumar, P., Paresh, K., and Narayan, G. (2011). Avocado fruit (*Persea americana* Mill) exhibits chemo-protective potentiality against cyclophosphamide induced genotoxicity in human Lymphocyte culture. *Journal of Experimental Therapeutics and Oncology*, 9: 221-230.

Rani, V., Deep, G., Singh, R. K., Palle, K., and Yadav, U. C. S. (2016). Oxidative stress and metabolic disorders: pathogenesis and therapeutic strategies. *Life Sciences,* 148: 183-193.

Ras, R. T., Hiemstra, H., Lin, Y., Vermeer, M. A., Duchateau, G. S. M. J. E. and Trautwein, E. A. (2013). Consumption of plant sterol-enriched foods and effects on plasma plant sterol concentrations - A meta-analysis of randomized controlled studies. *Atherosclerosis.* 230: 336-346.

Reczek, C. R., and Chandel, N. S. (2018). ROS Promotes Cancer Cell Survival through Calcium Signaling. *Cancer Cell*, 33(6): 949-951.

Rodríguez-Carpena, J. G., Morcuende, D., Andrade, M. J., Kylli, P. and Estévez, M. (2011). Avocado (*Persea americana* Mill.) phenolics, *in vitro* antioxidant and antimicrobial activities, and inhibition of lipid and protein oxidation in porcine patties. *Journal of Agricultural and Food Chemistry,* 59(10): 5625-5635.

Rodriguez-Sanchez, D. G., Flores-García, M., Silva-Platas, C., Rizzo, S., Torre-Amione, G., De la Peña-Diaz, A., et al., (2015). Isolation and chemical identification of lipid derivatives from avocado (*Persea americana*) pulp with antiplatelet and antithrombotic activities. *Food & Function*, 6(1), pp. 193-203.

Rodríguez-Sánchez, D. G., Silva-Platas, C., Rojo, R. P., García, N., Cisneros-Zevallos, L., García-Rivas, G., Hernández-Brenes, C. (2013). Activity-guided identification of acetogenins as novel lipophilic antioxidants present in avocado pulp (*Persea americana*). *Journal of Chromatography B,* 942–943: 37-45.

Rosenblat, G., Meretski, S., Segal, J., Tarshis, M., Schroeder, A., Zanin-Zhorov, A., Lion, G., Ingber, A., Hochberg, M. (2011). Polyhydroxylated fatty alcohols derived from avocado suppress in X ammatory response and provide non-sunscreen protection against UV-induced damage in skin cells. *Archives of Dermatological Research*, 303(4): 239-246.

Roth, G. (2009). Mannoheptulose glycolytic inhibitor and novel calorie restriction mimetic. *Experimental Biology*. Abstract n. 553.1. New Orleans, LA.

Ruiz, H. A., Rodríguez-Jasso, R. M., Fernandes, B. D., Vicente, A. A., and Teixeira, J. A. (2013). Hydrothermal processing, as an alternative for upgrading agriculture residues and marine biomass according to the biorefinery concept: A review. *Renewable and Sustainable EnergyReviews,* 21: 35-51.

Saavedra, J., Córdova, A., Navarro, R., Díaz-Calderón, P., Fuentealba, C., Astudillo-Castro, C., Toledo, L., Enrione, J. and Galvez, L. (2017). Industrial avocado waste: Functional compounds preservation by convective drying process. *Journal of Food Engineering*, 198: 81-90.

Salgado, J. M., Bin, C., Mansi, D. N. and Souza, A. (2008). Efeito do abacate (*Persea americana* Mill) variedade hass na lipidemia de ratos hipercolesterolêmicos. *Food Science and Technology*, 28(4): 922-928.

Sansano, S., Rivas, A., Pina-Pérez, M. C., Martinez, A., and Rodrigo, D. (2017). Stevia rebaudiana bertoni effect on the hemolytic potential of Listeria monocytogenes. *International Journal of Food Microbiology*, 250: 7-11.

Santos, M. A. Z., Alicieo, T. V. R., Pereira, C. M. P., Ramis-Ramos, G., and Mendonca, C. R. B. (2014). Profile of bioactive compounds in avocado pulp oil: influence of dehydration temperature and extraction method. *Journal of the American Oil Chemical Society*, 91: 19-27.

Segovia, F. J., Corral-Pérez, J. J., Almajano, M. P. (2016). Avocado seed: Modeling extraction of bioactive compounds. *Industrial Crops and Products*, 85: 213-220.

Shehata, M., and Soltan, S. S. (2013). Effects of bioactive component of kiwi fruit and avocado (fruit and seed) on hypercholesterolemic rats. *World Journal of Dairy & Food Sciences*, 8: 82-93.

Soldera-Silva, A., Seyfried, M., Campestrini, L. H., Zawadzki-Baggio, S. M., Minho, A. P., Molento, B. and Maurer, J. B. B. (2018). Assessment of anthelmintic activity and bio-guided chemical analysis of *Persea americana* seed extracts. *Veterinary Parasitology*, 251: 4-43.

Stefano, V., Avellone, G., Bongiorno, D., Indelicato, S., Massentib, R., and Bianco, R. (2017). Quantitative evaluation of the phenolic profile in fruits of six avocado (*Persea americana*) cultivars by ultra-high-performance liquid chromatography-heated electrospray-mass spectrometry. *International Journal of Food Properties*, 20(6): 1302-1312.

Stücker, M., Memmel, U., Hoffmann, M., Hartung, J., and Altmeyer, P. (2001). Vitamin B12 cream containing avocado oil in the therapy of plaque psoriasis. *Pharmacology and Treatment*, 203(2): 141-147.

Tabeshpour, J., Razavi, B. M., and Hosseinzadeh, H. (2017). Effects of avocado (*Persea americana*) on metabolic syndrome: A comprehensive systematic review. *Phytotherapy Research*. 31(6): 819-837.

Tango, J. S., Carvalho, C. R. L. and Soares N. B. (2004). Caracterização física e química de frutos de abacate visando a seu potencial para extração de óleo. *Revista Brasileira de Fruticultura, Jaboticabal - SP*, 26(1): 17-23.

Tesfay, S. Bertling, I. and Bower, J. (2011). Effects of postharvest potassium silicate application on phenolics and other anti-oxidant systems aligned to avocado fruit quality. *Postharvest Biology and Technology,* 60: 92-99.

Thongchai, W. (2014). Determination of retinol and alpha-tocopherol in avocado nanoemulsion with monolithic column. *The Thai Journal of Pharmaceutical Sciences*, 38(1): 1-56.

Tiwari, A. K. (2004). Antioxidants: new-generation therapeutic base for treatment of polygenic disorders. *Current Science*, 86(8): 1092-1102.

Tremocoldi, M., Rosalen, P., Franchin, M., Massarioli, A., Denny, C., Daiuto, C., Paschoal, J., Melo, P., and Alencar, S. (2018). Exploration of avocado by-products as natural sources of bioactive compounds. *Plos One*, 13(2): 1-12.

Uchenna, U. E., Shorib, A. B., and Baba, A. S. (2017). Inclusion of avocado (*Persea americana*) seeds in the diet to improve carbohydrate and lipid metabolism in rats. *Revista Argentina de Americanalogia y Metabolismo*, 4(3):140-148.

Ukwe, C. V., and Nwafor, S. V. (2004). Anti-ulcer activity of aqueous leaf extract of *Persea americana* (family-Lauraceae). *Nigerian Journal of Pharmaceutical Research*, 3(1): 91-95.

Ulbricht. C. E. (2016). An evidence-based systematic review of beta-sitosterol, sitosterol (22,23dihydrostigmasterol, 24-ethylcholesterol). *Journal of Dietary Supplements*, 13(1):35-92.

USDA (U.S. Department of Agriculture). (2011). Avocado, almond, pistachio, and walnut Composition. Nutrient Data Laboratory. *USDA National Nutrient Database for Standard Reference, Release 24*. U.S. Department of Agriculture. Washington, DC.

Vinha, A. F., Moreira, J., and Barreira, S. V. P. (2013). Physicochemical parameters, phytochemical composition, and antioxidant activity of the algarvian avocado (*Persea americana* Mill.). *Journal of Agricultural Science,* 5: 100-109.

Wang, L., Bordi, P. L., Fleming, J. A., Hill, A. M., and Kris-Etherton, P. M. (2015). Effect of a moderate fat diet with and without avocados on lipoprotein particle number, size, and subclasses in overweight and obese adults: A randomized, controlled trial. *Journal of the American Heart Association*, 4(1): e001355.

Wang, M., Zheng, W., Khuong, T., and. Lovatt, C. J. (2016). Developmental differences in antioxidant compounds and systems in normal and small-phenotype fruit of 'Hass' avocado (*Persea americana* Mill.). *Scientia Horticulturae*, 206:15-23.

Wang, W., Bostic, T. R., and Gu, L. (2010). Antioxidant capacities, procyanidins, and pigments in avocados of different strains and cultivars. *Food Chemistry*, 122: 1193-1198.

Wang, Y., Hodge, A. M., Wluka, A. E., English, D. R., Giles, G. G., O'Sullivan, R., Forbes, A., Cicuttini, F. M. (2007). Effect of antioxidants on knee cartilage and bone in healthy, middle-aged subjects: a cross-sectional study. *Arthritis Research & Therapy*, 9(4): R66.

Whiley, A.W., Schaffer, B., and Wolstenholme, B.N. (2002). *The avocado botany, production, and uses*, CABI Publishing, New York.

Yamassaki, F. T., Campestrini, L. H., Zawadzki-Baggio, S. F., and Maurer, J. B. B. (2017). Avocado leaves: Influence of drying process, thermal incubation, and storage conditions on preservation of polyphenolic compounds and antioxidant activity. *International Journal of Food Properties,* 20(S2): S2280-S2293.

Zhang, J., Zhu, W.-F., Xu, J., Kitdamrongtham, W., Manosroi, A., Manosroi, J., Tokuda, H., Abe, M., Akihisa, T. and Feng, F. (2018). Potential cancer chemopreventive and anticancer constituents from the fruits of Ficus hispida L.f. (Moraceae). *Journal of Ethnopharmacology*, 214: 37-46.

Zhou, L., Tey, C. W., Bingol, G., and Bi, J. (2016). Effect of microwave treatment on enzyme inactivation and quality change of defatted avocado puree during storage. *Innovative Food Science and Emerging Technologies,* 37: 61-67.

In: Avocado: Consumption and Health ISBN: 978-1-53618-204-0
Editors: M. G. Melgar Lalanne et al. © 2020 Nova Science Publishers, Inc.

Chapter 2

AVOCADO (*PERSEA AMERICANA MILL.*) POTENTIAL AS SOURCE OF BIOACTIVE COMPOUNDS FOR HUMAN HEALTH AND WELL-BEING

Nicolas Espitia, María Paula Diaz, Karol T. Medina, Sonia Lizcano and Javier A. Davila*

Engineering Department, Chemical Engineering Program,
University of Bogotá Jorge Tadeo Lozano, Bogotá, Colombia

ABSTRACT

Avocado (*Persea Americana Mill.*) has a promising potential as a source of valuable compounds due to its chemical composition. Thus, the content in oil, phenolic compounds, carotenoids, flavonoids, vitamins, antioxidants and anthocyanins among others, present not only in the pulp but also in the peel and seeds, makes avocado attractive for the extraction of bioactive compounds to be used as functional food or nutraceuticals with applications in chemical, food and pharmaceutical industries.

* Corresponding Author's Email: davila2344@yahoo.com.

Avocado oil pulp is one of the most attractive products to be extracted because this oil contains monounsaturated fatty acids, which are characteristics of a healthy diet. The applications of these compounds are constrained for the extractions and concentration yields because the technologies are not well established yet. Conserving the chemical properties of the valuable compounds from avocado is a factor in the extraction process. In order to obtain extracts free of solvents, it is necessary to optimize extraction processes as well as enhance concentrations so it can use this product in medicinal applications for human health and well-being.

Keywords: avocado, oil, bio compounds, extraction

1. Overview

This chapter presents the potential of valuable compounds from avocado fruit and the applications for human health and well-being. The attractive chemical composition of the avocado parts like the peel, seed, and pulp shows the value properties that can be added to products derived from this fruit, such as oil, anthocyanins and phenolic compounds with an antioxidant activity that are emphasized for human consumption.

The location, production, and varieties of avocados can define the specific uses of this fruit. As an illustration, the peel and seed composition are presented as a source of bioactive compounds for several applications for human health, while the oil composition shows the different uses from the medicinal field to industry practices.

2. Chemical Composition of Avocado (*Persea americana Mill.*)

2.1. Types and Location of Avocado and Its Common Uses

Persea Americana Mill or Avocado originates with the Aztecs cultivating this tropical and subtropical crop. However, this fruit has the advantage of growing in different types of weather from mild cold through warmer climates. The only circumstance it cannot thrive is in a desert environment (Tremocoldi et al. 2018).

Regions such as South America and Central America began harvesting *Persea Americana mill* in 1833, extending the plantain to Florida, one of the states with higher production in the United States. (Kendİr and Köroğlu 2018).

This fruit comes from the *Persea Americana Mill* family that has nearly 100 species of tropical trees, which can reach 18 meters height, including their branches. The characteristics can diverge based on the temperature, crop location, and seasons. Every seasonal fruit has a representative characteristic, such as color or texture. An illustration of these characteristics is the features fall and winter fruits, which can be soft and green.

The *Lauraceae Family* has a lifespan between 40 to 75 years with an amount of 50 fruits per tree depending on the crop, pest type, water supply, and orchard size, which makes them a significant investment in the food industry (Kendİr and Köroğlu 2018). To obtain a productive orchard, it is necessary to have an appropriate irrigation system with a plentiful water supply that can take care of the orchard size using the right system that can be sprinkler or drip (Mccarthy 2018).

Annually, an avocado orchard needs 5 ML per ha (Perriberton) to 21 ML/ha (Camaron). It can be made with different sprinklers along the rows between some trees, and they are usually positioned between every tree. So, the water will be running around at the right place, with the specific nutrients that the fruit needs, especially when the weather is hot. The Canopy system changes the way to deliver the water using an underground system that supplies the water requirements in the orchard as well. (Mccarthy 2018).

One of the main factors to take into account in this product is the temperature and the environment, which makes Mexico the best country for this fruit production with the perfect temperature (Chel-Guerrero et al. 2016). Additionally, this fruit can withstand temperatures up to - 4°C (Abaide et al. 2017).

Over the years, studies have shown that temperature higher than 30ºC is not healthy for an avocado crop, exposing an optimal temperature between 19ºC to a maximum of 24ºC (Olaeta 2003). Furthermore, other factors that can improve or decrease the avocado production are the rainfall and humidity. Countries like Mexico and Guatemala have rainfall between 60 to 150 mm and relative humidity of 40 to 70%. So, they have the best conditions for the manufacturer. An optimal production target would be a temperature up to 29ºC, rainfall from 1,100 mm to 1,500 mm, and relative humidity of 70-90% (Araújo et al. 2018). This type of fluctuation on a plantation can generate problems in the growth of avocado flowers, reduction in the roots and fruit acidity, and all these alterations can give a deterioration of the fruit or reduce the production yield.

The avocado/alligator pear or vegetable pear has the particularity to be heterogeneous, and these diversities are classified by the flowering pattern that is two, type A or B with the difference of pollination between morning and afternoon (Litz et al., 2007, (Mickelbart et al. 2012)). The *Persea Americana mill* can be classified into three big groups labeled on the respective origins and the specific physiognomies of the fruit. The Mexican (*Persea Americana var. drymifolia*) has a large seed size, the Guatemalan (*Persea nubigena var. guatemalensis*) possess a medium skin thickness, and the West Indian (*Persea Americana var. Americana*) develops an intense green color. Now, it is easy to find more than 500 types of avocado due to the climatic condition. However, the most popular is 'Hassco' and 'Fuerte' avocado. These two are the most significant fruits because their structure has high amounts of proteins and minerals. The variety of this fruit variation depends on the origin of the fruit, as is illustrated in Table 1 (Araújo et al. 2018).

This fruit has become one of the biggest exports in countries such as Mexico, Colombia, Brazil, Unites States, Israel and Dominican Republic (Mickelbart et al. 2012; Tremocoldi et al. 2018) the principal importers from different parts of the world include: France, Canada and Japan (Dávila et al. 2017). As of 2008, Western Australia started to develop *Persea Americana mill* orchards with optimal conditions controlling the temperature and water quality, improving from 18,575 tones in 2008 to 35,000 tones in 2018. (Mccarthy 2018).

Table 1. Varieties of Avocado (Araújo et al., 2018)

Variety	Parentage	Seed Size	Skin Color	Skin Thickness	Origin
Hass	Hybrid	Medium	Green	Medium	California
Bacon	Mexican	Large	Green	Thin	California
Reed	Guatemalan	Large	Green	Medium	California
Fuerte	Hybrid	Large	Green	Medium	Puebla, Mexico
Lam Hass	Hybrid	Medium	Black	Medium	California
Stewart	Mexican	Small	Black	Thin	California
McDonald	Guatemalan	Medium	Green/black	Thick	Hawaii
Nabal	Guatemalan	Large	Green	Thin	Guatemala
Dickey	Guatemalan	Small	Green	N/A	Mexico
Lewis	West Indian	Medium	Green	N/A	Hawaii
Puebla	Mexican	Large	Black	Thick	Mexico

2.2. Common Uses

Persea Americana Mill is one of the most significant products of the present. Their high nutritional content and the health benefits that this fruit has can be used in food industries, yielding products such as avocado tea, avocado vinegar, guacamole, avocado oil, fruit juice (Araújo et al. 2018). In the cosmetic industry, the *Persea Americana mill* can be used in creams, hair oil, medicine masks, and repair treatments (Dávila et al. 2017; Araújo et al. 2018) due to the polyphenols that this fruit develops in the chemical structure.

Another product that stands out is the starch that comes from the avocado seed, which has great potential in fields like pharmaceutical products, biodegradable polymers for food (jellies, candies, sausages or bread products) because it has a minor value of gelatinization temperature compared with the maize starch (Chel-Guerrero et al. 2016).

In the Agro Corporation, the *Persea Americana mill's* wastes have the potential to develop a product since it is easy to obtain them. Subsequently, some products can be achieved based on these agro-industrial wastes and reduce the environmental impact (Dávila et al. 2017).

2.3. Valuable Compounds and Applications from the Pulp of Avocado (Chemical Composition, Oil Content, Applications, Other Valuable Compounds)

The *Persea Americana mill* has an extraordinary nutrition scale with a low caloric level, making it part of the food group known as "superfood," which has garnered massive attention due to the ratio of nutrients compared with the number of calories. The average avocado shows that 1 oz. has only 50 calories divided into vitamins, minerals, phytonutrients, and unsaturated fatty acids (Tremocoldi et al. 2018). This fruit's structure is composed of minerals such as potassium, phosphorus, and vitamins like β-Carotenes that make it a valuable product with high impact around the world (Araújo et al. 2018). One of the most significant properties is the amount of phenolic acid, flavone, procyanidins, and organic hydrocinnamic acid, which makes it one of the best options on the market to have a healthy life and helping disease such as diabetes (Tremocoldi et al. 2018).

The *Persea Americana Mill* contains antioxidants such as *α-tocopherol* (Ashton et al. 2006) with values between 70 to 190 mg/kg of oil and non-lipid constituents present in the oil include chlorophylls ranging from 11 to 19 mg/Kg of oil (Wong et al. 2008a) and carotenoids from1.0 to 3.5 mg/kg of oil (Ashton et al. 2006; Tremocoldi et al. 2018).

The avocado pulp has higher amounts of insoluble and soluble fiber with a percentage ranging from 30 to 70%, fat-soluble nutrients like acid oleic and palmitoleic, this fruit has phytochemicals and sugars up to 70% and low quantities of acids such as linoleic (Araújo et al. 2018). The avocado pulp contains 228 kcal with 67% to 78% of lipids distributed throughout 12 - 24% of carbohydrates that can be around 8640 mg/100g, 1960 mg/100g of protein, 1600 mg/100 g of ash and 6800 mg/100g of fiber exposing the oil with polar lipids, glycolipids, phospholipids, and monounsaturated fatty acids. The mineral content is divided from 507 mg/100 g of potassium, 20 mg/100g of magnesium, eight g/100 g of sodium, 13 mg/100g of calcium, 54 mg/100g of phosphorus and vitamins like vitamin C with 8.8 mg/100g, 0.08 mg/100g of thiamine, 0.00197 mg/100g of Vitamin E or alpha-tocopherol, β -0.087 mg/100g of carotene, 1.91 mg/100g of niacin (Araújo et al. 2018).

Three big fragments are distributed in the fruit: exocarp, mesocarp, endocarp, which concentrate on the kernel that possesses only 2% of oil. The mesocarp is the part of the avocado that has the most considerable amount of oil, with values from 10 through 30% of oil. (Araújo et al. 2018). To get an optimal outcome of avocado oil, the extraction has to be mechanic and works similarly to the olive oil extraction; taking out the seed and peel, the pulp is decomposed at 50ºC for 60 minutes and presents two phases oil-water which are divided using a decanted and splitting the water with centrifuges. Both peel and seed wastes are sent to the orchards for soil conditioning or used as animal feed. This type of oil is indispensable to have healthy fruits, which does not have a physiological injury to get a low percentage of free fatty acids (Dávila et al. 2017).

The best avocado oil corresponds to the Hass because it can develop monounsaturated fatty acids with a high temperature, reaching up to 250ºC, which makes it perfect for frying oil. This oil is extracted in a cold press with a green color attributed to chlorophylls levels and carotenoids that comes from the mesocarp. In this process, it is fundamental to be meticulous when it comes to extraction, especially with the carotenoids since this has lutein and chlorophyll, which can produce oxidation. Another species that can generate this type of oil is 'Fuerte' but develops phytosterols, b-sitosterol with ranges from 2.23 to 4.48 mg/g oil that gives benefit to the commercial industry, being classified as the healthiest oil on the food market (Winners et al. 2010).

Another primary product in the food industry is guacamole, which is the conversion of the avocado into a batter combined with more ingredients such as onions, tomatoes, jalapenos, chili, garlic, sugars, salts, sodium, sodium alginate, disodium dehydrogenate phosphate, ascorbic and citric acid. The pulp requires a pH level between 6.7 and 5.2 to inhibit any kind of pathogen that can grow during this process. Additionally, it is essential to keep this product under -18ºC for at least 38 weeks to reduce the reproduction of *L. monocytogenes*, which is the principal pathogen (Iturriaga, Arvizu-Medrano, And Escartín 2002).

2.4. Valuable Compounds and Applications from Peel and Seed of the Avocado

The avocado seed is 15 or 16% of the fruit's weight of fatty acid, polyphenols, cyanogenic steroids, antioxidants, and potassium in the structure (Barbosa-Martin et al., 2016). The avocado seed also encloses saponin, flavonoids, phenols, cyanogenic glycosides phytosterols, triterpenes, and furanoic acids (Dávila et al. 2017). Multiple studies show the number of carbohydrates, minerals, and lipids that the *Persea Americana Mill* can have in their structure.

The seeds are usually oblate, conical or ovoid and are between 5 to 6.4 cm, hard, heavy and long, they have a thin shell, which is the endocarp that surrounds the core fruit, that can be used in the extraction of avocado oil such because it represents 21 or 30% the wastes of the *Persea americana* mill (Araújo et al. 2018; Tremocoldi et al. 2018). The seed is a great starch source due to the high amounts of polysaccharides that make up 30% of their structure; therefore, the avocado seed and peel can generate antimicrobial agents, antioxidants, or bioethanol.

Persea americana mill possesses pigments such as carotenoid and chlorophyll, which can be yellow, black, green or medium green. Most of them possess lutein, R-carotene - carotene, neoxanthin, chlorophylls a and b with large concentrations in order to obtain this pigments it is necessary to use in solvents such as diethyl ether, hexane, and acetone for an analytical method or methanol, acetonitrile, and ethyl acetate for HPLC method (Ashton et al. 2006).

3. Bioactive Compounds for Human Health

3.1. Avocado Oil Composition and Its Benefits for Human Health

3.1.1. Fatty Acid Profile

The palmitoleic, palmitic, linoleic, and oleic acids are those fatty acids found in the highest concentrations in avocado oil. The concentration of those fatty acids may vary among other researchers due to the extraction method that has been used. Table 2 shows the composition of fatty acids with different extraction methods.

Unsaturated fatty acids such as oleic and linoleic are essential for an optimal diet because they can delay or reduce damage to brain functions that cause diseases such as Parkinson's and Alzheimer's (Youdim, Martin, and Joseph 2000). Another health benefit is that they reduce the level of cholesterol in the blood since they help to increase the concentrations of HDL cholesterol (high-density lipoproteins) and regulate LDL levels (Szponar L, Mojska H 2008). Therefore, unsaturated fatty acids can decrease the development of cardiovascular diseases such as coronary heart disease and atherosclerosis (Bojarowicz and Woźniak 2008). Unsaturated acids also have anti-allergic and anti-inflammatory effects by increasing the level of the body's immune barrier (Kelley 2001). Another important aspect is the anticancer effect by reducing the development of nascent tumors and by inhibiting the proliferation of cancerous tissue (Ziemlański 2001).

In general, ω-6 fatty acids can be used in the treatment of diseases such as diabetes and obesity (Díaz-Gerevini et al. 2019), whereas ω-3 acids help in the development of the vision and brains of children in the fetal and postnatal period and also in the prevention of diseases in old age (Bianconi et al. 2018). However, the excess in the diet of ω-6 acids can induce the development of degenerative diseases and disorders in the functioning of the brain (Kwak et al. 2011), and an increase in the proportion of ω-6 and ω-3 acids increases the risk of obesity (Simopoulos 2016).

Table 2. Fatty acid profile of avocado oil with different extraction methods

Fatty acid	Soxhlet[1]	Ultrasound in water bath[1]	Ultra-Turrax[1] treatment[1]	Supercritical fluid[1]	Microwave[2]	Malaxation[3]
C14:0 Myristic acid	0.34 ± 0.01	ND	0.55 ± 0.38	1.58 ± 0.02	ND	ND
C16:0 Palmitic acid	24.33 ± 0.08	23.91 ± 0.34	24.22 ± 0.55	25.29 ± 0.97	19.5 ± 0.009	19.4 ± 1.3
C16:1 Palmitoleic acid	13.01 ± 0.17	13.42 ± 0.16	13.17 ± 0.15	13.08 ± 1.48	6.82 ± 0.05	9.0 ± 0.5
C18:0 Stearic acid	ND	ND	ND	12.80 ± 0.84	0.77 ± 0.04	0.6 ± 0.1
C18:1 Oleic acid	59.00 ± 0.57	58.19 ± 2.05	59.45 ± 0.95	50.35 ± 1.03	62.8 ± 0.32	51.6 ± 1.2
C18:2ω6 Linoleic acid	ND	ND	ND	ND	8.73 ± 0.24	18.0 ± 2
C18:3ω6 Linolenic acid	ND	ND	ND	ND	1.33 ± 0.00	1.2 ± 0.2

[1](Reddy, Moodley, and Jonnalagadda 2012).

[2](Guzmán-Gerónimo and Dorantes 2008),

[3](RAMÍREZ-ANAYA et al. 2018).

3.1.2. Bioactive Compounds

Avocado oil also contains bioactive compounds such as phytosterols (See Table 3), which can be found as esters with fatty acids, as free form, as glycosides or acylated steryl glycosides (Mo et al. 2013). The main phytosterols present in avocado oil are lupeol, β-Sitosterol, and stigmasterol, a diet rich in phytosterols can reduce the level of cholesterol in the blood according to some authors (Lottenberg et al. 2002), since phytosterols inhibit the intestinal absorption of cholesterol and act on total cholesterol and LDL cholesterol and also increases HDL levels.

Phytosterols have antifungal, antitumor, anti-inflammatory, and antibacterial actants. They have also been studied to heal wounds and therapeutic efficacy against diseases such as diabetes, arthritis, kidney and cardiovascular diseases.

Table 3. Concentrations of phytosterols in avocado oil (Mo et al. 2013)

Phytosterol	Concentration (μg/g)
β-Sitosterol	2629 ± 92
Campesterol	404.3 ± 8.9
Cycloartenol	347.1 ± 1.6
Lupeol	18.1 ± 2.3
Stigmasterol	172.6 ± 5.6

The β-Sitosterol helps strengthen the immune system because it increases the proliferation of lymphocytes and decreases the invading microorganisms by increasing the activity of natural killer cells (Paniagua-Perez et al. 2008). Also, this phytosterol helps reduce the accumulation of fat in the abdomen and compulsive consumption, therefore, it helps to lose weight.

The Lupeol demonstrated antiviral activity when tested against Herpes simplex virus 1 (HSV1) inhibiting 100% of viral plaque formation, and also when it was proven to inhibit the growth of influenza A virus (Tanaka et al. 2004; Parvez et al. 2018). It was found to be effective against the pathogenic protozoa that cause malaria, inhibiting parasite proliferation (Siddique and Saleem 2011). The Lupeol also has chemotherapeutic and chemopreventive potential; it has been reported to inhibit the growth of various types of tumors and particularly that it has no toxic effect on human cells at the doses at which it kills cancer cells (Saleem 2009). It was demonstrated in animal models that it inhibits diabetes since lupeol inhibits α-amylase and α-glucosidase (Saeed and Ahmed 2015; Ali, Houghton, and Soumyanath 2006).

The stigmasterol was studied in mice with skin cancer and was shown to have chemopreventive activity and antigenotoxic properties due to the decreased size of the tumor and the number of papillomas present in the mouse (Ali et al. 2015). It has also been reported the presence of lycopersene in avocado oil extracted with GLP and CO_2 at high pressures (Abaide et al. 2017), the lycopersene is a carotenoid that has been shown to have acaricidal activity (Ebadollahi et al. 2015).

3.2. Anthocyanins Rich Extracts from Avocado

Avocado changes color from green to purple-black when ripe (Cox et al. 2004) characterized the pigments during the ripening of "Hass" avocado and found that the total anthocyanins in the skin of the avocado increased as the fruit matured and identified the cyanidin 3-O-glucoside. Cox et al. (2004) have identified cyanidin-3-galactoside and cyanidin-3,5-diglucoside-p-coumarate as the anthocyanins responsible for the purple color during the ripening of the fruit.

The anthocyanins have several health benefits thanks to its antioxidant action, among the beneficial effects is the decrease in lipid level, the prevention of heart and cardiovascular diseases, anticancer, anti-inflammatory, and anti-allergic activity inhibits the activity of α-glucosidase and improves hyperglycemia (Motohashi 2012).

Anthocyanins have shown an essential role in the prevention of cancer by influencing the activity of enzymes, which are known for their detoxifying and antioxidant properties (Warner 2014). It has been shown that a diet supplemented with fruits rich in anthocyanins increases the levels of GSH and glutathione-S-transferase (Boateng et al. 2007), both play an essential role in the detoxification and protection of oxidative cell damage (Warner 2014). Anthocyanins have also shown an inhibitory effect on the growth of cancer cells (Hakimuddin, Paliyath, and Meckling 2004) reported a selective antiproliferative effect by observing the average growth of normal cells, while breast cancer cells (MCF-7) were inhibited, they suggest that these cells had become apoptotic.

The ability of anthocyanins to prevent cholesterol oxidation and the prevention of atherosclerosis has been demonstrated (Pascual-Teresa and Sanchez-Ballesta 2008). Also, researchers such as (Jayaprakasam et al. 2005) suggest that anthocyanins can combat risk factors for diseases such as type 2 diabetes, thanks to the ability of 3-O-glucoside to promote insulin secretion (Tsuda et al. 2005) demonstrated that this same anthocyanin significantly increases lipolytic activity.

Another beneficial effect for health is its potential to combat gout disease because the anthocyanins are capable of inhibiting xanthine oxidase which is an enzyme that plays a vital role in the catabolism of purines such as uric acid and urate which in high concentrations produces gout disease (Lee, Lim, and Choung 2013).

4. Industrial Production and Extraction of Valuable Compounds from Avocado

Valuable avocado compounds were presented in the previous sections of the chapter; next, an analysis of the technologies used for extracting valuable compounds with the respective operating conditions and yields is developed. Oil extraction requires the oil emulsion present in the pulp of the fruit as well as the breaking of the oil cells (Bizimana, Breene, and Csallany 1993). Among the most used technologies are the extraction of oil by cold pressing, extraction with solvents and extraction with supercritical fluids; also after the extraction of the oil by one of the technologies mentioned above, it proceeds to the refining the oil to increase its value, it is also necessary for the use of the oil in cosmetic products or food (Kurlaender 2008).

The production of avocado products began in 1964 in California with the freezing of the fruit for further processing into guacamole, and from this were obtained other products for cosmetic and food use, which are the main sectors that make use of avocado (Kurlaender 2008). In the case of avocado oil, it is stable and contains large amounts of oleic acid, which is desirable since it is a monounsaturated acid and makes it possible to use it for frying or to use it in salads. Moreover, in the case of cosmetic use, the oil has high penetration properties in the skin (Kurlaender 2008) that are present in a large number of cosmetic products.

4.1. Oil Extraction

Different technologies are used for avocado oil extraction. This section presents processes using cold pressing extraction, supercritical fluid extraction, and solvent extraction, which are technologies that allow oil extraction with different yields.

4.2. State of the Art about Oil Extraction

The research regarding the different methods of extraction of avocado oil has been described for several years, such as the method of enzymatic extraction described by Buenrostro and López (1986) were from the hypothesis that the oil is present inside the cells which are today a confirmed hypothesis. Hydrolysis was proposed with the use of a mixture of polygalacturonases, α amylase, and protease in order to obtain a high yield. Finally, the yields with this extraction were close to 74% of the oil contained in the fruit (Buenrostro and López-Munguia 1986).

Werman and Neeman (1987) evaluated the effects of pH, temperature and NaCl concentration during the centrifugal force process, determining the optimal conditions for this method 75°C, 5.5 of pH and a concentration of NaCl of 5%, where the highest acid values of 8.35 were obtained. Additionally, they found that an extraction process with ethanol allows a concentration of 192.9 ppm of chlorophyll that was the highest (Werman and Neeman 1987). During the investigations of avocado, Bizimana et al. (1993) develop an analysis of the different process conditions necessary for extraction, obtaining that the best conditions for the extraction of avocado oil are 5:1 water to avocado ratio, pH 5.5 and 5% of inorganic salt in solvent extraction, temperature between 75 - 98°C for recovery of 70 - 80% of avocado oil (Bizimana, Breene, and Csallany 1993).

After this type of research where conditions of operation were determined, it was preceded to combine extraction methods for increasing yields. Such as the case of Moreno et al. (2003) who from a comparison of 4 methods obtained that the highest yield of avocado was achieved with a microwave-hexane method and the method that hardly changed the properties of the avocado oil was microwave-squeezing, a method proposed in other studies (Ortiz Moreno et al. 2003). Regard to the microwave-squeezing method yields of 67% of the oil was obtained. Also, it was analyzed that the modifications of the oil were minimal and allowed affirming that this method allows for obtaining high-quality oil. Besides, a comparison was made with the extraction with acetone obtaining yields 12% (Ortiz Moreno et al. 2004).

Other works continue evaluating unconventional extraction methods concerning for the more common methods such as supercritical extraction (Corzzini et al. 2017) using supercritical CO_2 with pressures from 200 to 400 bar and temperatures between 40 - 80°C followed by extraction with CO_2/ethanol mixture (93:7 w/w) reporting recovery percentages of 98% (Corzzini et al. 2017). Other unconventional methods have been evaluated by Tan et al. (2018), who analyzed the extraction of avocado oil using subcritical CO_2 extraction and ultrasound-assisted aqueous extraction compared with solvent extraction, they determined that the ultrasound-assisted aqueous extraction has a high potential for the extraction due to the high percentages of fatty acids in the final product (Tan et al. 2018).

4.3. Cold Pressing for Oil Extraction

The technology of cold pressing is the vegetable oil extraction method traditionally used since 2006 (Ortiz Moreno et al. 2004). However, it has been replaced by other types of technologies. Cold-pressing extraction still being used due to the stability of oxidation of the oil with this technology (Bae and Lee 2008). Also, the process of extracting oil by cold pressing allows the product to be used for human consumption otherwise, as is the case with the method of solvent extraction that can only be used in the cosmetic industry (Bost, Smith, and Crane 2008).

The oil extracted by a cold-pressing process is called extra virgin avocado oil and within the properties present in the oil due to extraction by that method is the high level of pigments, chlorophyll, carotenoids, anthocyanins and in addition to these properties is characteristic of this oil the intense flavor and obtaining more significant benefits for human health (Bost, Smith, and Crane 2008) this because the extraction is done without the intervention of any substance that can cause adverse effects on the oil. Many of these properties are more present in the method of extraction with cold pressing that maintains the properties of the oil (Ding et al. 2007). Furthermore, because the extraction process by cold pressing allows increasing the properties of the oil, it also allows for the change of conventional oils to other plant-based ones, this would achieve the increase of positive effects at the agricultural level due to the decrease of erosion in the soil and pesticides, which is why it is believed that extraction by cold pressing generates high expectations worldwide (Bost, Smith, and Crane 2008).

The concentration of the substances within the oil depends on the maturity, the culture, and the storage of the same before the extraction process, as well as the method of extraction, used. Also, it is necessary to maintain a controlled environment during the storage of the avocado oil because there is a decrease in the concentration of chlorophyll that produces a change in the color of the oil from emerald green to yellow gold and decreases the quality of the oil (Wong et al. 2008b).

The process of extraction by cold pressing is developed at approximately 50°C. Furthermore, the extraction process can control the amount of free fatty acids and peroxide present in the product that should be below 0.8% w/w and 20 meq/kg, respectively (values for extra virgin olive oil) (Wong et al. 2008b). It was determined that even the avocado oil has higher properties than the olive oil although they have a similar extraction process, the values of free acids for the avocado oil are low than 0.5% w/w and peroxides values less than four meq/kg (Woolf et al. 2007).

Figure 1 shows the process flow diagram of an avocado oil extraction process by cold pressing. The process begins with a pretreatment consisting of a wash and then obtaining the pulp extracting the skin and seed of the fruit, the next step is to grind the pulp allowing to break the cells that contain the oil inside and release it, then mix it with water and proceed to decant and centrifuge at 3000-4000 rpm to obtain the extra virgin avocado oil finally. The cost of an avocado oil process plant using this technology is near to 1 million USD, and the equipment should be designed in stainless-steel (Woolf et al. 2007).

It has been reported that the performance of the extraction with cold pressing is 15.8% v/w. Also, it is an expensive process compared with other methods such as the thermo-mechanical processes (Dávila et al. 2017), but having a smoother pulp yield can increase because there is a reduction in the cell wall, which allows greater exposure of the oil after pressing, this without reaching the fruit. The fruit must be removed immediately from the production lot because it can significantly reduce the quality of the oil (Wong et al. 2008b).

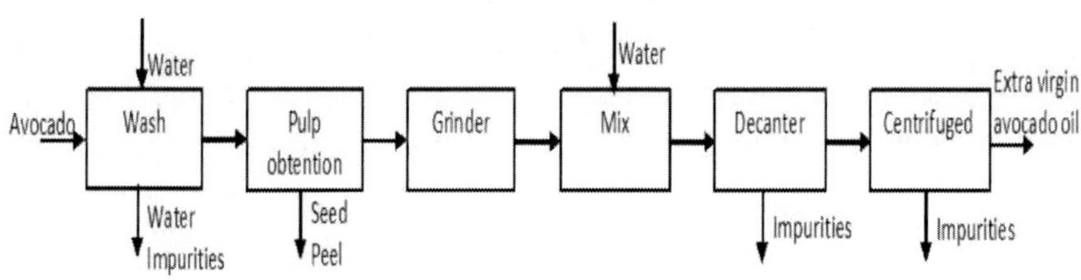

Figure 1. Block diagram flow of cold pressing avocado oil extraction.

It was determined that enzymatic inactivation occurs at 60°C, that is, the pulp must be dried at that temperature and then proceed to the extraction process, this process also allows the best preservation of the bioactive compounds of the oil and high values of activity antioxidant, as well as pigments such as chlorophyll and phenolic compounds for the solvent extraction method (Krumreich et al. 2018).

4.4. Supercritical Fluid Extraction

Supercritical fluids are currently used as solvents in a wide range of applications such as the extraction of bio compounds and vegetable oils, extraction of metal cations, polymer synthesis, among others. More than 90% of the extraction with supercritical fluid extraction (SFE) is carried out with carbon dioxide CO_2 since it has a significant advantage over its critical properties, as low pressure (74 bar) and temperature (32ºC). It is not toxic nor flammable, it is highly pure, and it is available in the market at affordable prices, after extraction, it is easy to remove CO_2 (Pourmortazavi and Hajimirsadeghi 2007).

By definition, a supercritical fluid is a substance that is at higher than its critical temperature and pressure, that is, above its critical point, a state in which there is no appreciable difference in the liquid and vapor aggregation states of said substance. At the critical point, the properties of the liquid and gaseous phase become very similar that they are practically indistinguishable, the fluid acquires a high diffusivity (typical of gases) and a high density (close to that of liquids, from 100 to 1000 times greater than the gases). Under these conditions, a supercritical fluid extraction has similar material transfer properties to those of gases and solvation characteristics like those of liquids. Also, the high diffusivity allows them to penetrate solid materials, and their density makes them capable of dissolving solutes. The compressible and, therefore, small changes in pressure can cause significant changes in density and solvent power. When a liquid-vapor mixture approaches the critical point fluctuations in the density of both phases, it begins to take place in regions of microscopic dimensions giving rise to a phenomenon of typical light dispersion known as "critical opalescence." Near to this critical point, small changes in the temperature and pressure of the substances result in considerable changes in properties such as density, which is why they have been attractive because this gives them remarkable abilities as solvents, which can also be adjusted to the measurement and necessity of a particular chemical process. Besides, they have excellent transport properties and very low surface tension. These attributes of a supercritical fluid have allowed many traditional industrial processes to innovate or optimize their production techniques, bringing substantial economic and environmental benefits (Tita, Cornejo, and Ambrogi 2010).

Below is an example of operation equipment for extraction with supercritical fluids (Figure 2). The equipment is fed from a cylinder (TK-1) containing carbon dioxide (generally with purity 99.8%), for the liquefaction of CO_2 a Chiller (H.E -1) is used, which will be recirculating a mixture of ethylene glycol with water that will lower the temperature of the CO_2 until it becomes liquid in the HE-1 heat exchanger. For the pressurization, a pump (P-1) is used, which operates with compressed air provided by a compressor. A heat exchanger (H.E-2) is used to increase the CO_2 temperature according to the extraction to be made. The container is surrounded by a heating band that is linked to the control system. The extraction is carried out in a high-pressure vessel (R-1) and the recovery of the extracts in a collector (R-2).

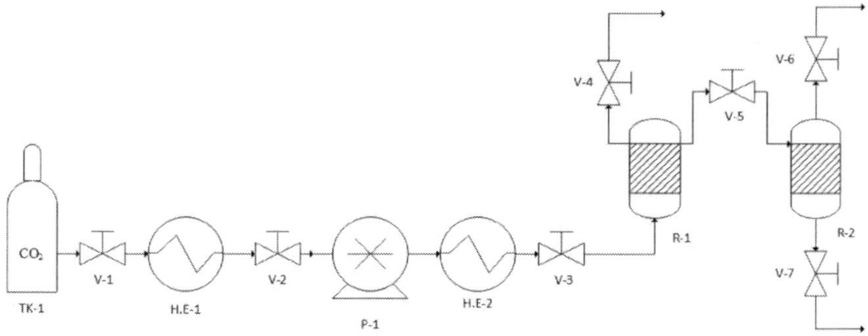

Figure 2. Process flow diagram for supercritical CO_2 extraction.

The solubility of vegetable oils in SC-CO_2 varies directly with temperature and pressure because CO_2 is quite compressible. The density, which is related to solute retention power, changes rapidly with changes at low pressures, while above 6,000 psi, the density does not change as quickly. Therefore, the expected increase in solubility with the increase in temperature would be observed at higher pressures, significantly improving extraction efficiency at a constant temperature. Where the extraction efficiency is defined in general terms as the weight of oil extracted per unit weight of CO_2 passed through the substrate on this basis (Friedrich and Pryde 1984).

There are many variables to be considered in the SFE. For this reason, computational methods have been developed in order to model the solubility of the analyte of interest in the plant matrix and carbon dioxide since the choice of extraction conditions is largely determined empirically. These models have been based on thermodynamics and transport phenomena such as mass and heat to estimate the behavior of supercritical fluids to the conditions to work (Pourmortazavi and Hajimirsadeghi 2007).

Other authors (Mageshni et al. 2012) have used SFE for obtaining oil from *Hass* and *Fuerte* avocado at conditions of 40 - 45°C, 420 - 450 bar, and 120 min, finding yields of 62.87 ± 0.29 and 59.56 ± 0.36 (g/100g), respectively. They found that SFE allows a broader range of fatty acids present in the extracted oil, just as the oils of the *Fuerte* variety had a higher monosaturated proportion of saturated fatty acids (3.45-3.70). Extraction by SFE produced better quality oil compared to the traditional Soxhlet extraction and with the lowest amount of oxidizing metals present. In this study, it was also possible to determine that the oils extracted with SFE are more suitable for products of the pharmaceutical and food industries since the compounds of interest with a high purity are obtained by not using solvents. They also propose that it is a feasible method in large-scale operations (Reddy, Moodley, and Jonnalagadda 2012).

Other authors (Botha 2003) carried out extractions of avocado oil with SFE during 2h with a CO_2 flow of 4.5 ml/min, the pressure between 354.6 and 539 bar, and temperatures between 37°C and 81°C obtaining yields close to 40%. They found that supercritical carbon dioxide was better than hexane as a solvent for avocado oil extraction since it avoids the extraction of chlorophyll that serves as a photosensitizer in oxidants. On the other hand, it found shorter extraction times, and with the absence of organic solvents, SFE is preferred to hexane extraction. They concluded that extractions with SFE avoid purification stages, which improves economic viability as an industrial process. Similarly, the concentration of unsaponifiable matter in avocado oil decreases as SFE extraction progresses (Botha and Mccrindle 2003).

4.5. Solvent Extraction of Oil

On the other hand, conventional methods of extracting oils with solvents such as Soxhlet technology are found, is the standard method of extracting solid samples. In addition to many methods of the EPA (U.S. Environmental Protection Agency) and the FDA (Food and Drugs Administration), use this classical technique as a proper method for the continuous extraction of solids. Its basis is in the separation based on the distribution between phases, the distribution of the solute between the two immiscible phases is a phenomenon of equilibrium that is described using the law of distribution, whose equilibrium constant is called distribution coefficient or the coefficient of distribution. The distribution constants allow us to calculate the solute concentration that remains in a solution after several extractions and allow them to determine the most efficient way to perform an extraction by extraction. Generally, a volatile solvent is used, which is heated until it reaches its state of evaporation, and when it condenses, it falls directly on the solid sample favoring the solubility of the system, which allows the dragging of oils, fats or certain substances. Some advantages of this technology are that the sample is repeatedly in contact with the solvent. Many pretreatments are not required after the extraction; it is a simple method to use, and good yields can be obtained. It allows the recovery of the solvent with broken evaporation. It should be noted that the choice of solvent and extraction time depends mainly on the nature of the analyte, cost of solvent, solubility, and polarity of the solid matrix and the solvent (Hernandez, Bojaca, and Duque 2014).

The extraction with organic solvents is the most common method for the extraction of different vegetable oils. To improve mass transfer, a pretreatment to the vegetable matrix should be enhanced where the avocado is cut, dried, and ground. Then the oil is extracted with organic solvents. The extraction capacity and quality of the avocado oil are affected by factors as the fruit ripeness and pulp moisture since it interferes with the effect of oil extraction so that the reduction of the water content in the pulp of the fruit is necessary before the extraction with solvent. To lower the moisture of the fruit pulp, there are two conventional drying approaches, oven drying, and cold drying. Generally, the solvents that include hexane and acetone are widely used for this type of oil extract from various matrices (Lozano et al. 1993). Moreno et al. 2003 reported that the extraction yield was 54% by extraction with hexane and 12% by extraction with acetone, respectively, using the Soxhlet method (AOAC 963.15), at the boiling temperatures of the respective solvents, the sample refluxed for 4h so that the extractor siphon 30 times to improve extraction (Ortiz Moreno et al. 2003).

The extraction of oil with solvents generates a remarkable difference in the cells of the plant matrix because it causes an irregular and rough form of idioblastic and parenchymal cells. Although a higher oil extraction yield can be obtained with an appropriate solvent, this technique has some drawbacks, such as environmental contamination and solvent residues in the final products, which limits the use of avocado oil in food and pharmaceutical applications (Qin and Zhong 2016).

CONCLUSION

The chemical composition of avocado reveals that all parts (peel, seed, and pulp) of this fruit have essential, valuable compounds that can be used for medicinal applications. An array of critical rich extracts can be obtained from avocado containing anthocyanins, vitamins, carotenoids, among others. Oil fraction from avocado also can be used for several diseases. Extraction techniques are of critical thinking in medicinal applications since the purity and quality of the extracts depend on the processes for extracting those compounds.

In light of the data present in this work, the whole avocado fruit has excellent potential for taking advantage of several valuable compounds. However, the technology and processes selected for the extraction and transformation of the different compounds are of importance since the quality and yield depend on the operational conditions as well as the final characteristics for human consumption purposes.

REFERENCES

Abaide, Ederson R., Giovani L. Zabot, Marcus V. Tres, Rafael F. Martins, Jean L. Fagundez, Lucielle F. Nunes, Susanne Druzian, et al. 2017. "Yield, Composition, and Antioxidant Activity of Avocado Pulp Oil Extracted by Pressurized Fluids." *Food and Bioproducts Processing* 102: 289–98. https://doi.org/10.1016/j.fbp.2017.01.008.

Ali, H, S. Dixit, D. Ali, S M Alqahtani, S. Alkahtani, and S. Alarifi. 2015. "Isolation and Evaluation of Anticancer Efficacy of Stigmasterol in a Mouse Model of DMBA-Induced Skin Carcinoma." *Drug Design, Development and Therapy* 9: 2793–2800. https://doi.org/10.2147/DDDT.S83514.

Ali, H, P. J. Houghton, and A. Soumyanath. 2006. "α-Amylase Inhibitory Activity of Some Malaysian Plants Used to Treat Diabetes; with Particular Reference to Phyllanthus Amarus." *Journal of Ethnopharmacology* 107 (3): 449–55. https://doi.org/10.1016/j.jep.2006.04.004.

Araújo, Rafael G, Rosa M. Rodriguez-Jasso, Héctor A. Ruiz, Maria Manuela E Pintado, and Cristóbal Noé Aguilar. 2018. "Avocado By-Products: Nutritional and Functional Properties." *Trends in Food Science & Technology* 80: 51–60. https://doi.org/https://doi.org/10.1016/j.tifs.2018.07.027.

Ashton, Ofelia B. O., Marie Wong, Tony K. McGhie, Rosheila Vather, Yan Wang, Cecilia Requejo-Jackman, Padmaja Ramankutty, and Allan B. Woolf. 2006. "Pigments in Avocado Tissue and Oil." *Journal of Agricultural and Food Chemistry* 54 (26): 10151–58. https://doi.org/10.1021/jf061809j.

Bae, E. K., and Sung Je Lee. 2008. "Microencapsulation of Avocado Oil by Spray Drying Using Whey Protein and Maltodextrin." *Journal of Microencapsulation* 25 (8): 549–60. https://doi.org/10.1080/02652040802075682.

Bianconi, S, G. Stutz, M. R. Solís, A. C. Martini, L M Vincenti, M F Ponzio, E Luque, C Avendaño, P Quiroga, and M E Santillán. 2018. "Maternal and Postnatal High-Fat Diets with High Ω6 : Ω3 Ratios Affect the Reproductive Performance of Male Offspring in the Mouse." *Reproduction, Fertility & Development* 30 (11): 1491–1502. https://doi.org/10.1071/RD17552.

Bizimana, V., W. M. Breene, and A. S. Csallany. 1993. "Avocado Oil Extraction with Appropriate Technology for Developing Countries." *Journal of the American Oil Chemists' Society* 70 (8): 821–22. https://doi.org/10.1007/BF02542610.

Boateng, J, M. Verghese, L Shackelford, L. T. Walker, J. Khatiwada, S Ogutu, D S Williams, et al. 2007. "Selected Fruits Reduce Azoxymethane (AOM)-Induced Aberrant Crypt Foci (ACF) in Fisher 344 Male Rats." *Food and Chemical Toxicology* 45 (5): 725–32. https://doi.org/https://doi.org/10.1016/j.fct.2006.10.019.

Bojarowicz, Halina, and BarbarA Woźniak. 2008. "Wielonienasycone Kwasy Tłuszczowe Oraz Ich Wpływ Na Skórke" ["Polyunsaturated Fatty Acids and Their Effect on the Skin"]. *Problemy Higieny i Epidemiologii* 89 (4): 471–75.

Bost, J.B, N Smith, and J Crane. 2008. "History, Distribution and Uses." In *The Avocado Botany, Production and Uses*, edited by R. Baxter, N. Hastings, A. Law, and E. J. Glass, second edition, 10–39. Boston: CABI.

Botha, B. M, and R. I. Mccrindle. 2003. "Supercritical Fluid Extraction of Avocado Oil," no. Ultrospec 3000: 11–13.

Buenrostro, M., and A. C. López-Munguia. 1986. "Enzymatic Extraction of Avocado Oil." *Biotechnology Letters* 8 (7): 505–6. https://doi.org/10.1007/BF01025210.

Chel-Guerrero, Luis, Enrique Barbosa-Martín, Agustino Martínez-Antonio, Edith González-Mondragón, and David Betancur-Ancona. 2016. "Some Physicochemical and Rheological Properties of Starch Isolated from Avocado Seeds." *International Journal of Biological Macromolecules* 86: 302–8. https://doi.org/10.1016/j.ijbiomac.2016.01.052.

Corzzini, Sayuri C. S., Helena D. F. Q. Barros, Renato Grimaldi, and Fernando A. Cabral. 2017. "Extraction of Edible Avocado Oil Using Supercritical CO_2 and a CO_2/Ethanol Mixture as Solvents." *Journal of Food Engineering* 194: 40–45. https://doi.org/10.1016/j.jfoodeng.2016.09.004.

Cox, Katy A, Tony K. McGhie, Anne White, and Allan B. Woolf. 2004. "Skin Colour and Pigment Changes during Ripening of 'Hass' Avocado Fruit." *Postharvest Biology and Technology* 31 (3): 287–94. https://doi.org/https://doi.org/10.1016/j.postharvbio.2003.09.008.

Dávila, Javier A., Moshe Rosenberg, Eulogio Castro, and Carlos A. Cardona. 2017. "A Model Biorefinery for Avocado (*Persea americana* Mill.) Processing." *Bioresource Technology* 243: 17–29. https://doi.org/10.1016/j.biortech.2017.06.063.

Díaz-Gerevini, Gustavo Tomás, Alejandro Daín, María Eugenia Pasqualini, Cristina B. López, Aldo R. Eynard, and Gastón Repossi. 2019. "Diabetic Encephalopathy: Beneficial Effects of Supplementation with Fatty Acids Ω3 and Nordihydroguaiaretic Acid in a Spontaneous Diabetes Rat Model." *Lipids in Health and Disease* 18 (1): 1–15. https://doi.org/10.1186/s12944-018-0938-7.

Ding, Haiming, Young Won Chin, A. Douglas Kinghorn, and Steven M. D'Ambrosio. 2007. "Chemopreventive Characteristics of Avocado Fruit." *Seminars in Cancer Biology* 17 (5): 386–94. https://doi.org/10.1016/j.semcancer.2007.04.003.

Ebadollahi, A, J. J. Sendi, A Aliakbar, and J. Razmjou. 2015. "Acaricidal Activities of Essential Oils from Satureja Hortensis (L.) and Teucrium

Polium (L.) against the Two Spotted Spider Mite, Tetranychus Urticae Koch (Acari: Tetranychidae).” *Egyptian Journal of Biological Pest Control* 25 (1): 171–76. https://www.scopus.com/inward/record.uri?eid=2-s2.0-84958239567&partnerID=40&md5=0dd689572e7e5eb4205442bf3069638e.

Friedrich, J. P., and E. H. Pryde. 1984. “Supercritical CO2extraction of Lipid-Bearing Materials and Characterization of the Products.” *Journal of the American Oil Chemists' Society* 61 (2): 223–28. https://doi.org/10.1007/BF02678773.

Guzmán-Gerónimo, Rosa I, and Lidia Dorantes. 2008. “Cambios En El Perfil de Ácidos Grasos y Microestructura de Aguacate Hass Tratado Con Microondas/Fatty Acids Profile and Microstructure of Avocado Puree after Microwave Heating.” *Archivos Latinoamericanos de Nutrición* 58 (3): 298–302. http://www.scielo.org.ve/scielo.php?script=sci_arttext&pid=S0004-06222008000300013&lng=en&tlng=en.

Hakimuddin, F, G. Paliyath, and K. Meckling. 2004. “Selective Cytotoxicity of a Red Grape Wine Flavonoid Fraction Against MCF-7 Cells.” *Breast Cancer Res Treat.* https://doi.org/https://doi.org/10.1023/B:BREA.0000021048.52430.c0.

Hernandez, Ana Maria, Rocio del Pilar Bojaca, and Marta E. Duque. 2014. “Determinación de Grasas y Aceites En Aguas Por El Método Soxhlet.” *Ideam*, 1–8. https://doi.org/10.11280/gee1973b.31.2951.

Iturriaga, Montserrat H., Sofia M. Arvizu-Medrano, and Eduardo F. Escartín. 2002. “Behavior of Listeria Monocytogenes in Avocado Pulp and Processed Guacamole.” *Journal of Food Protection* 65 (11): 1745–49. https://doi.org/10.4315/0362-028X-65.11.1745.

Jayaprakasam, Bolleddula, Shaiju K. Vareed, L. Karl Olson, and Muraleedharan G. Nair. 2005. “Insulin Secretion by Bioactive Anthocyanins and Anthocyanidins Present in Fruits.” *Journal of Agricultural and Food Chemistry* 53 (1): 28–31. https://doi.org/10.1021/jf049018+.

Kelley, Darshan, S., 2001. "Modulation of Human Immune and Inflammatory Responses by Dietary Fatty Acids." *Nutrition* 17 (7–8): 669–73.

Kendİr, Gülsen, and Ayşegül Köroğlu. 2018. *Evaluation of Avocado (Persea Americana Mill.) Leaves in Terms of Public Health* 22 (3): 347–56.

Krumreich, Fernanda D., Caroline D. Borges, Carla Rosane B. Mendonça, Cristina Jansen-Alves, and Rui C. Zambiazi. 2018. "Bioactive Compounds and Quality Parameters of Avocado Oil Obtained by Different Processes." *Food Chemistry* 257 (June 2017): 376–81. https://doi.org/10.1016/j.foodchem.2018.03.048.

Kurlaender, Azriel. 2008. "Avocados." In *Processing Fruits: Science and Technology*, edited by Diane Barret, Laszlo Somogyi, and Hosahalli Ramaswamy, second edition, 39:739–58. CRC Press.

Kwak, Jung Hyun, Jean Kyung Paik, Oh Yoen Kim, Yangsoo Jang, Sang-Hak Lee, Jose M Ordovas, and Jong Ho Lee. 2011. "FADS Gene Polymorphisms in Koreans: Association with Ω6 Polyunsaturated Fatty Acids in Serum Phospholipids, Lipid Peroxides, and Coronary Artery Disease." *Atherosclerosis* 214 (1): 94–100. https://doi.org/10.1016/j.atherosclerosis.2010.10.004.

Lee, Jin Hwan, Jung Dae Lim, and Myoung-Gun Choung. 2013. "Studies on the Anthocyanin Profile and Biological Properties from the Fruits of Acanthopanax Senticosus (Siberian Ginseng)." *Journal of Functional Foods* 5 (1): 380–88. https://doi.org/https://doi.org/10.1016/j.jff.2012.11.010.

Lottenberg, Ana M P, Valéria S Nunes, Edna R Nakandakare, Mônica Neves, Márcia Bernik, José E Santos, and Eder C R Quintão. 2002. "Plant Sterol Ester Efficiency on the Plasma Lipid Reduction in Moderate Hipercholesterolemic Subjects." *Arquivos Brasileiros de Cardiologia* 79 (2): 139–42.

Lozano, Y. F., C. Dhuique Mayer, C. Bannon, and E. M. Gaydou. 1993. "Unsaponifiable Matter, Total Sterol and Tocopherol Contents of Avocado Oil Varieties." *Journal of the American Oil Chemists' Society* 70 (6): 561–65. https://doi.org/10.1007/BF02545319.

Mccarthy, Alec. 2018. *Avocados Western Australia*. 2018. https://www.agric.wa.gov.au/avocados/avocados-western-australia-—-overview.

Mickelbart, Michael V., Paul W. Robinson, Guy Witney, and Mary Lu Arpaia. 2012. "'Hass' Avocado Tree Growth on Four Rootstocks in California. II. Shoot and Root Growth." *Scientia Horticulturae* 143: 205–10. https://doi.org/10.1016/j.scienta.2012.06.021.

Mo, Shunyan, Linlin Dong, W Jeffrey Hurst, and Richard B van Breemen. 2013. "Quantitative Analysis of Phytosterols in Edible Oils Using APCI Liquid Chromatography-Tandem Mass Spectrometry." *Lipids* 48 (9): 949–56. https://doi.org/10.1007/s11745-013-3813-3.

Motohashi, Noboru. 2012. *Anthocyanins : Structure, Biosynthesis and Health Benefits*. New York: Nova Science Publishers, Inc. http://ezproxy.utadeo.edu.co:2048/login?url=http://search.ebscohost.com/login.aspx?direct=true&db=e000xww&AN=541969&lang=es&site=eds-live&scope=site.

Olaeta, J. 2003. "Industrialización Del Aguacate: Estado Actual Y Perspectivas Futuras" ["Avocado Industrialization: Current Status and Future Prospects"]. *Actas V Congreso Mundial Del Aguacate*, 749–54.

Ortiz Moreno, Alicia, Lidia Dorantes, Juvencio Galíndez, and Elizabeth Cárdenas. 2004. "Effect of a Novel Oil Extraction Method on Avocado (*Persea americana* Mill) Pulp Microstructure." *Plant Foods for Human Nutrition* 59 (1): 11–14. https://doi.org/10.1007/s11130-004-0032-3.

Ortiz Moreno, Alicia, Lidia Dorantes, Juvencio Galíndez, and Rosa I. Guzmán. 2003. "Effect of Different Extraction Methods on Fatty Acids, Volatile Compounds, and Physical and Chemical Properties of Avocado (*Persea americana* Mill.) Oil." *Journal of Agricultural and Food Chemistry* 51 (8): 2216–21. https://doi.org/10.1021/jf0207934.

Paniagua-Perez, R, E. Madrigal-Bujaidar, S. Reyes-Cadena, I. Alvarez-Gonzalez, L. Sanchez-Chapul, J. Perez-Gallaga, N Hernandez, G. Flores-Mondragon, and O Velasco. 2008. "Cell Protection Induced by Beta-Sitosterol: Inhibition of Genotoxic Damage, Stimulation of Lymphocyte Production, and Determination of Its Antioxidant

Capacity." *Archives of Toxicology* 82 (9): 615–22. http://search.ebscohost.com/login.aspx?direct=true&db=edsgao&AN=edsgcl.229196320&site=eds-live.

Parvez, Mohammad K, Perwez Alam, Ahmed H Arbab, Mohammed S Al-Dosari, Tawfeq A Alhowiriny, and Saleh I Alqasoumi. 2018. "Analysis of Antioxidative and Antiviral Biomarkers β-Amyrin, β-Sitosterol, Lupeol, Ursolic Acid in Guiera Senegalensis Leaves Extract by Validated HPTLC Methods." *Saudi Pharmaceutical Journal* 26 (5): 685–93. https://doi.org/https://doi.org/10.1016/j.jsps.2018.02.022.

Pascual-Teresa, S, and M. T. Sanchez-Ballesta. 2008. "Phytochemistry Reviews". *Proceedings of the Phytochemical Society of Europe. Anthocyanins: From Plant to Health.* 7: 281–99.

Pourmortazavi, Seied Mahdi, and Seiedeh Somayyeh Hajimirsadeghi. 2007. "Supercritical Fluid Extraction in Plant Essential and Volatile Oil Analysis." *Journal of Chromatography A* 1163 (1–2): 2–24. https://doi.org/10.1016/j.chroma.2007.06.021.

Qin, Xiaoli, and Jinfeng Zhong. 2016. "A Review of Extraction Techniques for Avocado Oil." *Journal of Oleo Science* 65 (11): 881–88. https://doi.org/10.5650/jos.ess16063.

Ramírez-Anaya, Jessica del Pilar, Ariana Jocelyn MANZANO-Hernández, Ernesto Tapia-Campos, Karina Alarcón-Domínguez, and Ma Claudia Castañeda-Saucedo. 2018. "Influence of Temperature and Time during Malaxation on Fatty Acid Profile and Oxidation of Centrifuged Avocado Oil." *Food Science and Technology* 38 (2): 223–30. https://doi.org/10.1590/1678-457x.33116.

Reddy, Mageshni, Roshila Moodley, and Sreekanth B. Jonnalagadda. 2012. "Fatty Acid Profile and Elemental Content of Avocado (*Persea americana* Mill.) Oil -Effect of Extraction Methods." *Journal of Environmental Science and Health - Part B Pesticides, Food Contaminants, and Agricultural Wastes* 47 (6): 529–37. https://doi.org/10.1080/03601234.2012.665669.

Saeed, Ramsha, and Dildar Ahmed. 2015. "Bioactive Compounds from Carissa Opaca Roots and Xanthine Oxidase and Alpha-Amylase

Inhibitory Activities of Their Methanolic Extract and Its Fractions in Different Solvents." *Pharmacognosy Research* 7 (4): 295–301. https://doi.org/10.4103/0974-8490.158440.

Saleem, M., 2009. "Lupeol, a Novel Anti-Inflammatory and Anti-Cancer Dietary Triterpene." *Cancer Letters* 285 (2): 109–15. https://doi.org/10.1016/j.canlet.2009.04.033.

Siddique, Hifzur Rahman, and Mohammad Saleem. 2011. "Beneficial Health Effects of Lupeol Triterpene: A Review of Preclinical Studies." *Life Sciences* 88 (7): 285–93. https://doi.org/https://doi.org/10.1016/j.lfs.2010.11.020.

Simopoulos, P Artemis. 2016. "An Increase in the Omega-6/Omega-3 Fatty Acid Ratio Increases the Risk for Obesity." *Nutrients*. https://doi.org/10.3390/nu8030128.

Szponar L, Mojska H, Ołtarzewski MG. Tłuszcze. 2008. *Normy Żywienia Człowieka. Podstawy Prewencji Otyłości i Chorób Niezakaźnych* [*Human Nutrition Standards. Fundamentals of Obesity Prevention and Non-communicable Diseases*]. Edited by Barbara Bułhak-Jachymczyk Mirosław Jarosz. PZWL Medic. Warszawa.

Tan, Chin Xuan, Gun Hean Chong, Hazilawati Hamzah, and Hasanah Mohd Ghazali. 2018. "Comparison of Subcritical CO2and Ultrasound-Assisted Aqueous Methods with the Conventional Solvent Method in the Extraction of Avocado Oil." *Journal of Supercritical Fluids* 135 (November 2017): 45–51. https://doi.org/10.1016/j.supflu.2017.12.036.

Tanaka, Tomonori, Tsuyoshi Ikeda, Miho Kaku, Xing-Hua Zhu, Masafumi Okawa, Kazumi Yokomizo, Masaru Uyeda, and Toshihiro Nohara. 2004. "A New Lignan Glycoside and Phenylethanoid Glycosides from *Strobilanthes Cusia* B<small>REMEK</Small>." *Chemical and Pharmaceutical Bulletin* 52 (10): 1242–45. https://doi.org/10.1248/cpb.52.1242.

Tita, Gerardo, M. Cornejo, and a. Ambrogi. 2010. "Diseño y Construcción de Una Planta Piloto Multipropósito de CO2 Supercrítico" ["Design and Construction of a Supercritical CO2 Multipurpose Pilot Plant"]. *Cuadernos de La Facultad de Ingeniería e Informática*, no. 5: 46–55.

http://www.apiperu.com/argentina/trabajos/IPPIA_083_Tita_Gerardo.pdf.

Tremocoldi, Maria Augusta, Pedro Luiz Rosalen, Marcelo Franchin, Adna Prado Massarioli, Carina Denny, Érica Regina Daiuto, Jonas Augusto Rizzato Paschoal, Priscilla Siqueira Melo, and Severino Matias De Alencar. 2018. “Exploration of Avocado By-Products as Natural Sources of Bioactive Compounds.” *PLoS ONE* 13 (2): 1–13. https://doi.org/10.1371/journal.pone.0192577.

Tsuda, Takanori, Yuki Ueno, Hitoshi Kojo, Toshikazu Yoshikawa, and Toshihiko Osawa. 2005. “Gene Expression Profile of Isolated Rat Adipocytes Treated with Anthocyanins.” *Biochimica et Biophysica Acta (BBA) - Molecular and Cell Biology of Lipids* 1733 (2): 137–47. https://doi.org/https://doi.org/10.1016/j.bbalip.2004.12.014.

Warner, L. M., 2014. *Handbook of Anthocyanins: Food Sources, Chemical Applications, and Health Benefits*. Biochemistry Research Trends Series. Nova Science Publishers, Incorporated. https://books.google.com.co/books?id=sE74oAEACAAJ.

Werman, M. J., and I. Neeman. 1987. “Avocado Oil Production and Chemical Characteristics.” *Journal of the American Oil Chemists' Society* 64 (2): 229–32. https://doi.org/10.1007/BF02542007.

Winners, Aocs Award, President News, Latin American, and Section Ilsi. 2010. *What Is Unrefi Ned, Extra Virgin Cold-Pressed Avocado Oil ?*.

Wong, M., O. Ashton, C. Requejo-Jackman, T. McGhie, A. White, L. Eyres, N. Sherpa, and A. Woolf. 2008a. “Avocado Oil: The Color of Quality.” In *Color Quality of Fresh and Processed Foods*, 983:328–49. Washington, DC: American Chemical Society. https://doi.org/10.1021/bk-2008-0983.ch024.

———. 2008b. “Avocado Oil: The Color of Quality.” In *Color Quality of Fresh and Processed Foods*, 983:328–49. Washington, DC: American Chemical Society. https://doi.org/10.1021/bk-2008-0983.ch024.

Woolf, Allan, Marie Wong, Laurence Eyres, Tony McGhie, Cynthia Lund, Shane Olsson, Yan Wang, et al. 2007. “Avocado Oil.” In *Gourmet and Health-Promoting Specialty Oils*, 136:73–126.

Youdim, K. A, A. Martin, and J. A. Joseph. 2000. "Essential Fatty Acids and the Brain: Possible Health Implications." *International Journal of Developmental Neuroscience* 18 (4–5): 383–99. https://doi.org/10.1016/S0736-5748(00)00013-7.

Ziemlański, Światosław. 2001. *Normy Żywienia Człowieka: Fizjologiczne Podstawy* [*Human Nutrition Standards: Physiological Foundations*]. Wydaw. Lekarskie PZWL.

In: Avocado: Consumption and Health ISBN: 978-1-53618-204-0
Editors: M. G. Melgar Lalanne et al.

Chapter 3

NUTRACEUTICAL PROFILE OF AVOCADO AND HEALTH

Lidia Dorantes Alvarez*[1,*]*, Alicia Ortiz Moreno*[1]*,
***Felipe García Ochoa*[2] *and Nancy Ordaz Trinidad*[1]**
[1]Biochemical Engineering Department, National Polytechnic Institute, National School of Biological Sciences, Mexico City, Mexico
[2]Biophysics Department, National Polytechnic Institute, National School of Biological Sciences, Mexico City, Mexico

ABSTRACT

Avocado consumption has been associated with positive changes in the biochemical indicators of metabolic disorders, mainly caused by poor nutrition, aging, overweight, and obesity. Recently, avocado oil and extracts have shown anticancer activity and have contrasted the effects of diabetes and Osteoarthritis. The objective of the present work was to establish a non-exhaustive review about the evidence of the effects of avocado pulp, avocado oil, avocado extracts, and some isolated avocado compounds on improving health, in clinical or preclinical studies, and its

* Corresponding Author's Email: lidiadorantesa@gmail.com, ldoran@ipn.mx.

correlation to nutraceuticals present in avocado. Some examples are as follows: 1) avocado pulp and oil intake had beneficial effects on dyslipidemia and cardiovascular risk by reducing the serum levels of total cholesterol, total triacylglycerols, and low-density lipoproteins, most likely due to the high levels of phytosterols and monounsaturated fatty acids in avocados; 2) bioactive compounds extracted from avocado seeds and pulp have shown anticancer activity; 3) avocado soybean unsaponifiables reduced knee pain and cartilage damage in the treatment of Osteoarthritis; and 4) avocado oil decreases diastolic and systolic blood pressure in streptozotocin-induced diabetic rats. Some relevant avocado nutraceutical effects are as follows: i) acetogenins show antiplatelet and antithrombotic activities; ii) perseitol and D-mannoheptulose reduce insulin levels in serum, and iii) penstemide is reported to be anti-inflammatory. The limitation of some of the works is the lack of clinical research; however, most of these findings may be useful in the design of food supplements based on avocado nutraceuticals.

Keywords: avocado nutraceuticals, avocado, and obesity, avocado and health

1. INTRODUCTION

The increasing number of individuals worldwide with obesity, who are overweight, or who have diabetes and cancer (WHO), supports the idea of research on alternative solutions to these human diseases. Avocado consumption has been associated with the improvement of human health. Therefore, the objective of the present work is to elaborate a non-exhaustive review about the evidence of the effects of avocado pulp, avocado oil, avocado extracts and some isolated avocado compounds on improving health in clinical or preclinical studies in subjects (animal or human) with obesity, overweight, diabetes and cancer and its correlation to nutraceuticals present in avocado. The beneficial effects of avocado can be attributed to several biologically active compounds:

1. Phytosterols (β-sitosterol, stigmasterol, and campesterol), with β-sitosterol being the most predominant, compounds widely studied as anticholesterolemic agents

2. Polyphenols such as quinic and caffeoylquinic as well as chlorogenic acids, rutin, quercetin-diglucoside, and epicatechin
3. Monounsaturated fatty acids, mainly oleic acid
4. Carotenoids such as lutein and zeaxanthin
5. Unsaponifiable lipids

2. Evidence of Beneficial Effects of Avocado

2.1. Beneficial Effects of Avocado Product Intake in Metabolic Disorders

One of the most well-studied benefits of avocado intake is that it has favorable effects on indicators of metabolic disorders, such as the reduction of cholesterol. In their review article, Dreher and Daveport (2013) concluded that the majority of clinical studies showed that avocado consumption helps support cardiovascular health in overweight patients. Fonseca Duarte et al. (2016) emphasized the health benefits of avocado pulp and oil, especially in reducing cholesterol and preventing cardiovascular diseases, due to the compounds present in the lipidic fraction, such as 3 and 6 omega fatty acids, and a high content of oleic acid, phytosterols, tocopherols, and squalene.

A study (Haddad et al., 2018) showed that postprandial changes in $PYY_{3\text{-}36}$, GIP, and GLP-1 concentrations were associated with expected changes in the visual analog scale in overweight adults after avocado consumption. Also, decreased GLP-1 was observed in meals that included avocado, which contributed to increased satisfaction and a decreased desire to eat. Sabaté et al. (2015) reported improved insulin and glucose values, as well as increased satiety via GLP-1, in healthy overweight adults consuming Hass avocado fruits. This effect is attributed to the fat and dietary fiber of the avocado.

In the Adventist Health Study, the effect of avocado intake on weight and body mass index (BMI) was studied (Heskey et al., 2019). Avocado intake (standard serving size 32 g/day) was monitored by a food frequency questionnaire in individuals with a mean age of 56 years of US and Canada. They reported that avocado consumers had significantly lower weight (74.5 kg versus 77.9 kg) and body mass index (26.0 kg/m^2 versus 27.3 kg/m^2) than non-consumers.

The addition of 68 g of avocado to a high-fat meal (hamburger) was able to attenuate the postprandial vasoconstrictive effect of a high-fat diet and may reduce the postprandial proinflammatory and vaso-constrictory effects of a meat patty. Healthy participants who consumed 250 g of beef had an increase in serum IL-6, but this increase was attenuated when avocado was ingested. The practice of adding a slice of avocado to various products may have cardiovascular health benefits (Li et al. 2013).

Table 1. The beneficial effect of avocado in overweight and obesity. Dyslipidemia and cardiovascular risk improvement by the consumption of avocado products

Model	Type of study	Oil or Pulp	Doses	Beneficial effects	References
Human	26 healthy overweight subjects in randomized three × three crossover study	Pulp	~50 to 90 g according to the energy needs of the participant	Decreased GLP within three h Increased satisfaction and a decreased desire to eat	Haddad et al., 2018
Human	26 healthy overweight subjects in a randomized 3 x 3 crossover design	Pulp	½ of avocado A control group was included	Increased satisfaction by 26% and decreased the desire to eat by 40% over 3 hours	Wien et al., 2013
Human	Longitudinal analysis in vegetarians	Pulp	32 g	Consumption of avocados may reduce adult weight gain	Heskey et al., 2019
Human	Eleven healthy subjects	Pulp	68 g of avocado	Maintenance of IkBa, decreased activation of NFkB, no changes in IL-6 level	Li et al., 2013

Table 2. The beneficial effect of avocado products on dyslipidemia and atherogenic risk

Model	Type of study	Oil or Pulp	Doses	Beneficial effects	References
Rats	Sucrose-fed male Sprague-Dawley rats	Oil	7.5% of oil in the diet	Decreased TG, VLDL, and LDL levels	(Carvajal-Zarrabal et al. 2014)
Rats	Thirty-five male Wistar rats	Pulp reduced in calories	2 g/kg/day, for seven weeks	Reduced TC, LDL-C, and TG levels	(Pahua-Ramos et al. 2014)
Rats	Sprague-Dawley rats	Oil	900 mg/kg/day	Reduced LDL-C and TG levels, and HDL-C level was increased	(Tan et al., 2018)
Human	Thirteen healthy subjects with a diet containing saturated fats	Oil	50 g of fat (25 g monounsaturated fatty acids, mainly oleic acid, provided by avocado oil	Ameliorated postprandial levels of insulin, glucose, TC, LDL, TG, C-reactive protein and interleukin-6	Furlan et al., 2017
Human	26 healthy overweight subjects in randomized three × three crossover study	Pulp	~50 to 90 g according to the energy needs of the participant	Decreased GLP within three h Increased satisfaction and a decreased desire to eat	Haddad et al., 2018

The beneficial effects of avocado pulp and oil have been studied for dyslipidemia and cardiovascular risk (Table 2). Carvajal-Zarrabal et al. (2014) evaluated the effects of avocado oil administration on the biochemical markers of the cardiovascular risk profile in 250 rats divided into five groups: 1) one was fed a basic diet; 2) another group was fed a basic diet plus 30% sucrose solution 3) In addition to the sucrose diet, one of the groups was fed olive oil and 4) another group was fed centrifuged avocado oil and 5) avocado oil obtained by solvent extraction. Both avocado oils showed a similar effect to olive oil; both reduced the levels of cholesterol, triglycerides, VLDL, LDL, and the highly sensitive C-reactive protein in plasma, most likely to attenuate the inflammation process. The effect of reduced-calorie avocado paste (2 g/kg/day, for seven weeks) has been studied in rats fed a high-fructose diet. It showed a significant reduction in TC, LDL, TG, and increased insulin sensitivity (Pahua-Ramos et al., 2014).

The hypocholesterolemic activity of avocados in high-cholesterol diet-induced hypercholesterolemia rats was reported; once the hypercholesterolemia model was established, the rats continued the high-cholesterol diet while being orally administered avocado oil (450 and 900 mg/kg body weight per day) for four weeks. In the hypercholesterolemia control group, the administration of avocado oil increased the serum level of high-density lipoprotein-cholesterol (HDL-C). It significantly reduced the serum levels of total cholesterol (TC), total triacylglycerols (TG), and LDL-C at the end of the experiment. The researchers reported that the hypocholesterolemic action of avocado oil was due to its high levels of phytosterols and MUFA (Tan et al., 2018). These phytochemicals participate as antioxidant and anti-inflammatory compounds to diminish oxidative stress for endothelial dysfunction and lipoprotein oxidation in hypercholesterolemia.

Furlan et al. (2017) studied the effects of substituting avocado oil for butter in overweight subjects and reported the postprandial metabolic responses of avocado oil to the hypercaloric-hyperlipidemic diets of the overweight subjects. In this cohort study, overweight subjects consumed either a control meal (the diet consisted of butter, bacon, potatoes, eggs, iced sugar and wheat bread) or a test meal (the diet consisted of avocado oil, bacon, potatoes, eggs, iced sugar, and wheat bread), and blood collections were performed at defined times over a 240 min period. The results of their study indicated that the postprandial plasma levels of C-reactive proteins, glucose interleukin-6, insulin, TG, TC, and LDL-C of the subjects were improved in those who ate avocado oil along with a hypercaloric-hyperlipidic breakfast (eggs, bacon, wheat bread, potatoes, and reed sugar).

Campusano-Bublitz et al. (2016) studied the effects of avocado in hyperlipidemic mice induced with a high-fat diet over 28 days. Avocado pulp ingestion significantly reduced the total cholesterol, triglycerides, LDL, VLDL, and the index of atherogenic risk in the mice.

2.2. Consumption of Avocado Products and Diabetes

Mature avocado plants contain heptose D-mannoheptulose and its polyol form, perseitol. These compounds have various functions as antioxidants and energy sources, contributing approximately 6.0 g in 100 g (Tesfay et al., 2012). Preliminary D-mannoheptulose research suggests that the compounds may support blood glucose control and weight management (Roth, 2009).

Lima et al., 2012 investigated the hypoglycemic properties of avocado and observed that oral administration of a hydroalcoholic leaf extract (0.15 and 0.3 g/kg/day, for four weeks) to streptozotocin-induced diabetic rats reduced their blood glucose levels and improved their metabolic states. They proposed that the mechanism of this effect could be the regulation of glucose uptake in the liver and muscles through Akt/PKB activation and consequently restoring the intracellular energy balance.

Wien, Haddad, Oda, & Sabaté (2013) investigated 26 healthy overweight subjects in a randomized clinical trial and revealed that the consumption of half of Hass avocado significantly reduced the blood insulin and glucagon-like peptide-1 levels in the subjects. It was concluded that D-mannoheptulose might be responsible for the hypoglycemic effect of glycolysis decline via hexokinase inhibition and weight control via appetite reduction.

Ortiz-Avilla et al. (2015) reported the protective effects of commercial avocado oil on exacerbated oxidative stress and impaired electron transport chain function in the liver mitochondria of STZ-induced diabetic rats. The rats were orally administered avocado oil (1 mL/250 g body weight per day) for three months. At the end of the experiment, a reduction in lipid peroxidation, oxidative stress, and reactive oxygen species production in the liver mitochondria of the rats were observed. Furthermore, this study also demonstrated that avocado oil intake normalized the serum levels of

TC and TG caused by diabetes but failed to normalize hyperglycemia, polyphagia, and polydipsia. The findings of this study indicate the possibility of avocado oil to attenuate the negative physiological effects of diabetes on the oxidative condition of liver mitochondria. Furthermore, the potential of avocado oil intake to reduce insulin resistance in high-sucrose diet-induced diabetic rats was investigated by Toro-Equihua et al. (2016).

2.3. Anticancer Activity of Avocado Byproducts

The compounds present in avocado have been evaluated in cancer cell lines. Acetogenins are compounds found in the pulp and seed and are an essential group of fatty acid derivatives characterized by one unsaturated end and one highly oxygenated end. The seeds contain 5048.1 ± 575.5 mg/kg fresh weight of acetogenins, and the pulp contains 3107.0 ± 207.2 mg/kg fresh weight of (2S,4S)-1-acetoxy-2,4-dihydroxy-n-heptadeca-16-ene; Persediene, Persenone-C, Persenone-A, Persenone-B, and Persin have been identified in both the pulp and seeds, and AcO-avocadenyne has only been detected in the seeds (Pacheco et al., 2017; Rodriguez-Sanchez et al., 2015). An extract of the pulp, called D003, induces apoptosis in human oral cancer cells via regulation of reactive oxygen species at high concentrations (>30 μ/ml) and inhibits cell proliferation at lower concentrations (5-20 μg/ml) by a nonapoptotic mechanism. Two acetogenins inhibit the EGFR/RAS/RAF/MEK/ERK1/2 pathway at two different levels: inhibiting the activation of epidermal growth factor receptors (Tyr1173) and suppressing the activation of c-RAF (Ser 338) (D'Ambrosio, Han, Pan, Douglas Kinghorn, & Ding, 2011). Avocatin B is a mixture of avocadene and avocadyne, two 17-carbon polyhydroxylated fatty alcohols, and it possesses novel anticancer activity by selectively inducing apoptosis of leukemia and leukemia stem cells (Lee et al., 2015). This effect is due to the inhibition of fatty acid oxidation (which ignores the protection of leukemia cells against apoptosis), depletion of endogenous antioxidants (NADH NADPH, GSH), increase in whole-cell ROS levels, and triggered the mitochondrial release of cytochrome C and apoptosis-inducing factor (Tcheng, Samudio, Lee, Minden, & Spagnuolo, 2017). The content of avocadyne in the pulp and seeds was found to be 0.18 ± 0.04 mg/g dry weight and 0.41 ± 0.02 mg/g, respectively. Avocadene is found in pulp and seeds at 0.22 ± 0.04 mg/g and 0.43 ± 0.04 mg/g DW, respectively (Ahmed, Smith, Henao, Stark, & Spagnuolo, 2018).

2.4. Beneficial Effects of Avocado Soybean Unsaponifiables on the Treatment of Osteoarthritis

Avocado soybean unsaponifiables (ASUs) are compounds formed from soybean and avocado oil extracts (3:1). These compounds have been studied for the treatment of Osteoarthritis. Osteoarthritis is a chronic degenerative disease characterized by the degradation of articular cartilage, synovial inflammation, atypical bone growth, and possible damage to ligaments and menisci. Cytokines such as IL-1b and TNF-a influence pathways that promote disease progression. Treatment of ASUs in a cell model of LPS-activated chondrocytes decreased the IL-1b, TNF-a, COX-2, and iNOS levels (Au et al., 2007). A concentration of 10 mg/ml of ASUs in the chondrocyte cell line suppressed the translocation of the p50/p65 heterodimer, reduced the phosphorylation of ERK1/2 and decreased NF-κB activation, and the effect may be due to the presence of sterols that dysregulated the presence of reactive oxygen species (Gabay et al., 2008).

In an animal model, a mixture of the fractions unsaponifiables of avocado and soybean decreased the expression of iNOS and TNF-α in a monoiodoacetate-induced osteoarthritis model, and this result was also demonstrated by histopathological and ultrastructure studies to reduce the damage in cartilage subchondral bone tissue (Al-Afify et al., 2018). One of the primary markers of inflammation is that PGE2 inhibits the production of proteoglycans and induces the production of proinflammatory cytokines. A combination of avocado soybean unsaponifiables, rich in phytosterols, with lipoic acid, reduced prostaglandin E2 expression, and suppressed NF-κB translocation (Frondoza et al. 2018).

Table 3. The beneficial effect of avocado soybean unsaponifiables in osteoarthritis

Model	Type of study	Doses	Beneficial effects	References
Cell Culture	Articular chondrocytes	25 mg/ml	Decreased TNF-a, IL-1b, COX-2, and iNOS expression	Au et al., 2007
Rats	Male adult Sprague-Dawley rats	27.5 mg/kg/day (ASU capsules from Avosoya, DBK) for three weeks	Attenuated modifications in cartilage and synovial membranes. Reduced oxidative stress oxidative inflammation in cartilage and subchondral bone	Al-Afify et al., 2018
Cell culture	Articular chondrocytes	8.3 µg/Ml of ASU and 2.5 µg/mL of lipoic acid	Inhibited PGE_2 production, decreased translocation/activation of NF-B	Frondoza et al., 2018
Cell Culture	Chondrocytes	10 mg/ml	Decreased MMP-3, MMP-13, and PGE2 expression levels and suppressed p65 translocation	Gabay et al., 2008
Human	4186 patients with symptomatic knee osteoarthritis	300 mg ASU capsules for six months	Reduced pain scale of the knee	Głuszko et al., 2016
Human	Hip osteoarthritis patients	300 mg ASU capsules	Decreased joint space narrowing	Maheu et al., 2014
Human	Arthralgia and temporomandibular joint osteoarthritis	200 mg of soybean oil and 100 mg capsules	Reduced pain scale (visual analog scale)	Catunda et al., 2016

In humans, a study in Poland was carried out in knee osteoarthritis patients who were administered 300 mg ASU capsules for six months; the patients showed a decrease in the knee pain scale as well as the use of analgesics and nonsteroidal anti-inflammatory drugs (Głuszko et al., 2016). In another study, 189 hip osteoarthritis patients were treated daily with 300 mg of ASUs, and the joint space narrowing of the patients was reduced compared with that in patients treated with placebo upon the completion of the study (Maheu et al., 2014). Similar results were also reported in patients with arthralgia and temporomandibular joint Osteoarthritis. These patients showed pain scale improvements (visual analog scale) when they received 200 mg of soybean oil and 100 mg of avocado oil (Catunda et al., 2016).

Conclusion

It may be concluded that studies at the cellular level in animals and humans show that avocado oil, avocado pulp, and some extracts, have a positive impact in the treatment of hyperlipidemia, cardiovascular risk, obesity, diabetes, and Osteoarthritis. In some cases, these effects have been correlated with the nutraceutical profile of avocado. In the future, it is expected that more investigations through more clinical trials focus on the molecular structure of nutraceuticals of avocado and its correlation with the beneficial effects on human health.

Acknowledgment. The authors recognize the support of Instituto Politécnico Nacional SIP20201014.

References

Ahmed, N., Smith, R. W., Henao, J. J. A., Stark, K. D., & Spagnuolo, P. A. (2018). *Analytical Method to Detect and Quantify Avocatin B in Hass Avocado Seed and Pulp Matter*. 81 (4): 818-824. https://doi.org/http://doi.org/10.1021/acs.jnatprod.7b00914.

Al-Afify, Aliaa S. A., Gehan El-Akabawy, Neveen M. El-Sherif, Fatma El-Nabawya A. El-Safty, and Mostafa M. El-Habiby. (2018). Avocado Soybean Unsaponifiables Ameliorates Cartilage and Subchondral Bone Degeneration in Mono-Iodoacetate-Induced Knee Osteoarthritis in Rats. *Tissue and Cell* 52: 108-15. https://doi.org/https://doi.org/10.1016/j.tice.2018.05.001.

Au, R. Y., T K Al-Talib, A. Y. Au, P. V. Phan, and C. G. Frondoza. (2007). Avocado Soybean Unsaponifiables (ASU) Suppress TNF-α, IL-1β, COX-2, INOS Gene Expression, and Prostaglandin E_2 and

Nitric Oxide Production in Articular Chondrocytes and Monocyte/Macrophages. *Osteoarthritis and Cartilage* 15 (11): 1249-55. https://doi.org/10.1016/j.joca.2007.07.009.

Campuzano-Bublitz, Miguel A., Ruth A. Ubrán, Alma L. Rolón, Goretti E. M. Diarte, Carmen M. Coronel, and María L. Kennedy. (2016). Influence of Avocado (*Persea americana*) Pulp Consumption on Lipid Metabolism in Diet-induced Normolipidemic and Hyperlipidemic mice. *Latin American Nutrition Archives* 66 (4): 279-86.

Carvajal-Zarrabal, Octavio, Cirilo Nolasco-Hipolito, M. Guadalupe Guadalupe Aguilar-Uscanga, Guadalupe Melo-Santiesteban, Patricia M. M Hayward-Jones, and Dulce M. M Barradas-Dermitz. (2014). Avocado Oil Supplementation Modifies Cardiovascular Risk Profile Markers in a Rat Model of Sucrose-Induced Metabolic Changes. Edited by Fabrizia Bamonti. *Disease Markers* 2014: 386425. https://doi.org/10.1155/2014/386425.

Catunda, I. S., B. C. do E. Vasconcelos, E. S. de S. Andrade, and D. F. N. Costa. (2016). Clinical Effects of an Avocado-Soybean Unsaponifiable Extract on Arthralgia and Osteoarthritis of the Temporomandibular Joint: Preliminary Study. *International Journal of Oral and Maxillofacial Surgery* 45 (8): 1015-22. https://doi.org/https://doi.org/10.1016/j.ijom.2016.01.008.

D'Ambrosio, Steven M., Chunhua Han, Li Pan, A. Douglas Kinghorn, and Haiming Ding. (2011). Aliphatic Acetogenin Constituents of Avocado Fruits Inhibit Human Oral Cancer Cell Proliferation by Targeting the EGFR/RAS/RAF/MEK/ERK1/2 Pathway. *Biochemical and Biophysical Research Communications* 409 (3): 465-69. https://doi.org/https://doi.org/10.1016/j.bbrc.2011.05.027.

Dreher, Mark L., and Adrienne J. Davenport. (2013). Hass Avocado Composition and Potential Health Effects. *Critical Reviews in Food Science and Nutrition* 53 (7): 738-50. https://doi.org/10.1080/10408398.2011.556759.

Duarte, Patrícia Fonseca, Marcia Alves Chaves, Caroline Dellinghausen Borges, and Carla Rosane Barboza Mendonça. (2017). Avocado: Characteristics, Health Benefits, and Uses. *International News on Fats,*

Oils and Related Materials 28 (3): 28-32. https://doi.org/10.1590/0103-8478cr20141516.

Frondoza, Carmelita G., Lowella V. Fortuno, Mark W. Grzanna, Stacy L. Ownby, Angela Y. Au and Ann M. Rashmir-Raven. (2018). α-Lipoic acid Potentiates the Anti-Inflammatory Activity of Avocado/Soybean Unsaponifiables in Chondrocyte Cultures. SAGE Publications, *Cartilage* 9 (3): 304-312. https://doi.org/10.1177/1947603516686146.

Furlan, Cibele Priscila Busch, Sandra Costa Valle, Elin Östman, Mário Roberto Maróstica, and Juscelino Tovar. (2017). Inclusion of Hass Avocado-Oil Improves Postprandial Metabolic Responses to a Hypercaloric-Hyperlipidic Meal in Overweight Subjects. *Journal of Functional Foods* 38: 349-54. https://doi.org/https://doi.org/10.1016/j.jff.2017.09.019.

Gabay, O., M. Gosset, A. Levy, C. Salvat, C. Sanchez, A. Pigenet, A. Sautet, C. Jacques, and F. Berenbaum. (2008). Stress-Induced Signaling Pathways in Hyalin Chondrocytes: Inhibition by Avocado-Soybean Unsaponifiables (ASU). *Osteoarthritis and Cartilage* 16 (3): 373-84. https://doi.org/https://doi.org/10.1016/j.joca.2007.06.016.

Głuszko, Piotr, and Małgorzata Stasiek. (2016). Symptom-Modifying Effects of Oral Avocado/Soybean Unsaponifiables in Routine Treatment of Knee Osteoarthritis in Poland. An Open, Prospective Observational Study of Patients Adherent to a 6-Month Treatment. *Reumatologia* 54 (5): 217-26. https://doi.org/10.5114/reum.2016.63661.

Haddad, Ella, Michelle Wien, Keiji Oda, and Joan Sabaté. (2018). Postprandial Gut Hormone Responses to Hass Avocado Meals and Their Association with Visual Analog Scores in Overweight Adults: A Randomized three × 3 Crossover Trial. *Eating Behaviors* 31: 35-40. https://doi.org/https://doi.org/10.1016/j.eatbeh.2018.08.001.

Heskey, Celine, Keiji Oda, and Joan Sabaté. (2019). Avocado Intake, and Longitudinal Weight and Body Mass Index Changes in an Adult Cohort. *Nutrients* 11 (3). https://doi.org/10.3390/nu11030691.

Lee, Eric A, Leonard Angka, Sarah-Grace Rota, Thomas Hanlon, Andrew Mitchell, Rose Hurren, Xiao Ming Wang, et al. (2015). "Targeting

Mitochondria with Avocatin B Induces Selective Leukemia Cell Death." *Cancer Research* 75 (12): 2478 LP-2488. https://doi.org/10.1158/0008-5472.CAN-14-2676.

Li, Zhaoping, Angela Wong, Susanne M. M Henning, Yanjun Zhang, Alexis Jones, Alona Zerlin, Gail Thames, Susan Bowerman, Chi-Hong Tseng, and David Heber. (2013). Hass Avocado Modulates Postprandial Vascular Reactivity and Postprandial Inflammatory Responses to a Hamburger Meal in Healthy Volunteers. *Food & Function* 4 (3): 384-91. https://doi.org/10.1039/C2FO30226H.

Lima, C. R., C. F. B., Vasconcelos, J. H. Costa-Silva, C. A. Maranhão, J. Costa, T. M. Batista, E. M. Carneiro, L. A. L. Soares, F. Ferreira, and A. G. Wanderley. (2012). Anti-Diabetic Activity of Extract from *Persea americana* Mill. Leaf via the Activation of Protein Kinase B (PKB/Akt) in Streptozotocin-Induced Diabetic Rats. *Journal of Ethnopharmacology* 141 (1): 517-25. https://doi.org/https://doi.org/10.1016/j.jep.2012.03.026.

Maheu, Emmanuel, Christian Cadet, Marc Marty, Dominique Moyse, Isabelle Kerloch, Philippe Coste, Maxime Dougados, et al. (2014). Randomised, Controlled Trial of Avocado-Soybean Unsaponifiable (Piascledine) Effect on Structure Modification in Hip Osteoarthritis: The ERADIAS Study. *Annals of the Rheumatic Diseases* 73 (2): 376 LP-384. https://doi.org/10.1136/annrheumdis-2012-202485.

Ortiz-Avila, Omar, Marco Alonso Gallegos-Corona, Luis Alberto Sánchez-Briones, Elizabeth Calderón-Cortés, Rocío Montoya-Pérez, Alain R. Rodriguez-Orozco, Jesús Campos-García, Alfredo Saavedra-Molina, Ricardo Mejía-Zepeda, and Christian Cortés-Rojo. (2015). Protective Effects of Dietary Avocado Oil on Impaired Electron Transport Chain Function and Exacerbated Oxidative Stress in Liver Mitochondria from Diabetic Rats. *Journal of Bioenergetics and Biomembranes* 47 (4): 337-53. https://doi.org/10.1007/s10863-015-9614-z.

Pacheco, Adriana, Dariana Graciela Rodríguez-Sánchez, Raúl Villarreal-Lara, Jesús Manuel Navarro-Silva, Carolina Senés-Guerrero, and Carmen Hernández-Brenes. (2017). Stability of the Antimicrobial Activity of Acetogenins from Avocado Seed, under Common Food

Processing Conditions, against Clostridium Sporogenes Vegetative Cell Growth and Endospore Germination. *International Journal of Food Science & Technology* 52 (11): 2311-23. https://doi.org/10.1111/ijfs.13513.

Pahua-Ramos, María Elena, Leticia Garduño-Siciliano, Lidia Dorantes-Alvarez, German Chamorro-Cevallos, Julieta Herrera-Martínez, Obed Osorio-Esquivel, and Alicia Ortiz-Moreno. (2014). Reduced-Calorie Avocado Paste Attenuates Metabolic Factors Associated with a Hypercholesterolemic-High Fructose Diet in Rats. *Plant Foods for Human Nutrition* 69 (1): 18-24. https://doi.org/10.1007/s11130-013-0395-4.

Rodriguez-Sanchez, Dariana Graciela, Mirthala Flores-García, Christian Silva-Platas, Sheryl Rizzo, Guillermo Torre-Amione, Aurora De la Peña-Diaz, Carmen Hernández-Brenes, and Gerardo García-Rivas. (2015). Isolation and Chemical Identification of Lipid Derivatives from Avocado (*Persea americana*) Pulp with Antiplatelet and Antithrombotic Activities. *Food & Function* 6 (1): 192-202. https://doi.org/10.1039/C4FO00610K.

Roth, George, Mike Hayek, Stefan Massimino, Gary Davenport, Robert Arking, Andrzej Bartke, Michael Bonkowski, and Don Ingram. (2009). Mannoheptulose: Glycolytic Inhibitor and Novel Caloric Restriction Mimetic. *The FASEB Journal* 23 (1_supplement): 553.1-553.1. https://doi.org/10.1096/fasebj.23.1_supplement.553.1.

Sabaté, J., M. Wien, and E. Haddad. (2015). Post-Ingestive Effects of Avocados in Meals on Satiety and Gastric Hormone Blood Levels. *Human Health Nut.,* 459-461.

Tan, Chin Xuan, Gun Hean Chong, Hazilawati Hamzah, and Hasanah Mohd Ghazali. (2018). Hypocholesterolemic and Hepatoprotective Effects of Virgin Avocado Oil in Diet-Induced Hypercholesterolaemia Rats. *International Journal of Food Science & Technology* 53 (12): 2706-13. https://doi.org/10.1111/ijfs.13880.

Tcheng, Matthew, Ismael Samudio, Eric A. Lee, Mark D. Minden, and Paul A. Spagnuolo. (2017). The Mitochondria Target Drug Avocatin B Synergizes with Induction Chemotherapeutics to Induce Leukemia

Cell Death. *Leukemia & Lymphoma* 58 (4): 986-88. https://doi.org/10.1080/10428194.2016.1218005.

Tesfay, S. Z., I. Bertling, and J. P. Bower. (2012). D-Mannoheptulose and Perseitol in 'Hass' Avocado: Metabolism in Seed and Mesocarp Tissue. *South African Journal of Botany* 79: 159-65. https://doi.org/https://doi.org/10.1016/j.sajb.2011.10.006.

Toro-Equihua, Mario Del, Raymundo Velasco-Rodríguez, Raúl López-Ascencio, and Clemente Vásquez. 2016. Effect of an Avocado Oil-Enhanced Diet (*Persea Americana*) on Sucrose-Induced Insulin Resistance in Wistar Rats. *Journal of Food and Drug Analysis* 24 (2): 350-57. https://doi.org/10.1016/j.jfda.2015.11.005.

Wien, Michelle, Ella Haddad, Keiji Oda, and Joan Sabaté. (2013). A Randomized 3 x 3 Crossover Study to Evaluate the Effect of Hass Avocado Intake on Post-Ingestive Satiety, Glucose and Insulin Levels, and Subsequent Energy Intake in Overweight Adults. *Nutrition Journal* 12 (1): 155. https://doi.org/10.1186/1475-2891-12-155.

In: Avocado: Consumption and Health ISBN: 978-1-53618-204-0
Editors: M. G. Melgar Lalanne et al.

Chapter 4

OXIDATIVE STABILITY AND SHELF LIFE OF AVOCADO OIL

***Alan Javier Hernández-Álvarez*[1], *Oscar Lopez-Franco*[2] and *Guiomar Melgar-Lalanne*[3,*]**
[1]School of Food Science and Nutrition, University of Leeds, Leeds, UK
[2]Instituto de Ciencias de la Salud
[3]Instituto de Ciencias Básicas, Universidad Veracruzana,
Xalapa, Veracruz, Mexico

ABSTRACT

Avocado oil (AO) is rich in monounsaturated fatty acids (more than 70%) with lower levels of acidity and oxidation than virgin olive oil. The natural presence of tocopherols and polyphenols act as antioxidants and prevents oxidation. However, the level of oxidation is affected by the chemical composition of the fatty acid fraction (presence of free fatty acids, saturated, and unsaturated fat ratio), the presence of antioxidant compounds (tocopherols, polyphenols, and chlorophylls), as well as the extraction technologies employed and the storage conditions. The shelf

[*] Corresponding Author's Email: gmelgar@uv.mx.

life of AO is an indicator of the final quality of the product, and it is relevant for preserving its high value functional and sensorial quality. It is necessary to protect the oil against lipid oxidation, which can have many deleterious effects on the quality, including the formation of unpleasant compounds and the oxidation of functional secondary metabolites. To reduce the oxidation rate during the storage, certain factors such as the presence of light, oxygen, certain metals, and the storage temperature should be appropriately controlled. In this review, the different technologies applied to reduce oxidation and to improve the shelf life and final quality of AO are summarized and discussed, including avocado cultivars, ripeness degree, extraction, and storage.

Keywords: avocado oil, shelf life, lipid oxidation, fatty acid profile, storage

1. INTRODUCTION

Avocado oil (AO) is an edible oil obtained from the fruits of the avocado tree (*Persea americana* Mill.), native to Central America, which currently grows in tropical and subtropical climates across the world (Flores et al., 2019). The AO is highly appreciated for culinary purposes like salad dressing, sautéing, roasting, frying, and searing for its high smoke point (250°C). Moreover, avocado oil, especially cold-pressed (virgin and extra virgin oil), is widely appreciated for its functional properties and potential health benefits for its significant amount of oleic acid (Wong et al., 2011). Avocado world trade has been intensified during the last decades, mostly in fresh fruits, which currently means close to 90% of the total market, due to the high international price of the fruit. Therefore, only damaged fruits and lesser quality are used to produce avocado oil. However, the growing interest in avocado fruit and oil is given for its potential health benefits, according to the World Health Organization (WHO) (Bollinedi et al., 2020).

The avocado fruit pulp is extremely rich in oleic acid (close to 70%), a monounsaturated fatty acid (MUFA) with substantial health benefits reported. Additionally, the fatty fraction is also rich in secondary but important bioactive metabolites such as vitamins, phytosterols, chlorophylls, carotenoids, and tocopherols, which may have a positive effect against some metabolic diseases and even may enhance the biological absorption of some carotenoids (Sun-Waterhouse et al., 2011; Unlu et al., 2005). Thus, AO has a high omega 6/omega-3 ratio, close to 14.05 due to the high amount of linoleic acid (omega 6), and a PUFA/SFA ratio of 0.81 (Berasategui et al., 2012; Flores et al., 2019). Finally, the AO, mainly when it is extracted by cold pressing, is rich in antioxidants like carotenoids, tocopherols, and polyphenols (Cerecedo-Cruz et al., 2018), which are close to the ones found in olive oil.

The consumption of rich MUFA oils is preferable to the rich PUFA oils, both at the technological and the health-promoting level since MUFA are more stable to oxidation. However, few oils that can be found in the market are rich in MUFA; among them, olive oil is the most known and the most consumed worldwide. Although, nowadays, rapeseed oil, hazelnut oil, avocado oil, among others, can also be found in the gourmet sections (Visoli et al., 2020). MUFA is growing interest for their potential biological effects in the prevention and treatment of many diseases related to the diet and lifestyle (Vingering et al., 2010). In this fraction, the oleic acid is especially essential for its demonstrated benefits, which include a cardioprotective effect and the ability to reduce the low-density lipoproteins of cholesterol and the ability to reduce the blood pressure (Visoli et al., 2020).

Nevertheless, oleic acid is not an essential fatty acid and can be synthesized in humans by the stearoyl-CoA desaturase 1 (SCD1). This enzyme catalyzes the biosynthesis of MUFA from SFA, either derived from the diet or synthesized de novo preferable from palmitic and stearic acids, although, some metabolic diseases may limit its presence (Piccinin et al., 2019). Technologically, the high MUFA presence in the AO gives this oil higher stability to oxidation. However, the high presence of chlorophylls, mostly in cold-pressed oils, could act as photo pro-oxidants when the oil is exposed to the light. Moreover, the AO has a medium level of natural antioxidants like tocopherols, polyphenols, and carotenoids. Thus, it is vital to have proper storage conditions avoiding light to increase the shelf life of the oil.

The rancidification (lipid oxidation) is the main quality problem detected in fats and oils, and it causes unpleasant taste and odor resulting from the complete or incomplete oxidation or hydrolysis of fats and oils. Time, temperature, light, air, exposed surface, moisture, nitrogenous organic material, and traces of metals are factors responsible for rancidity. Therefore, the extraction conditions, the storage, and the culinary process affect the final quality and oxidation of food rich in fats and oils (Campanella et al., 1999). To avoid the lipid oxidation of oils, different strategies have been proposed, from the optimization of extraction and storage conditions (low temperature, low moisture, darkness, and use of inert atmospheres) to the added antioxidants from natural or artificial sources.

In this review, the different technologies applied to reduce oxidation and to improve the shelf life and final quality of AO are summarized and discussed.

2. Avocado Oil Composition and Physicochemical Characteristics

The lipidic fraction of avocado fruit constitutes between 49.5 to 65.29 g/100 dry mass, depending on the variety (Jorge et al., 2015). However, the oil extraction yield is conditioned by the extraction method used rather than the variety. The extraction oil yield is higher when the solvent extraction by Soxhlet is used, although the final quality of the oil is worst since the final oil must be refined, and most of the antioxidant compounds are eliminated. Obtaining oil by cold pressing could improve its extraction yield by slightly increasing the process at 60°C (Krumreich et al., 2018), and with the previous lyophilization of the fruits (dos Santos et al., 2014).

Nevertheless, the fatty fraction of avocado depends basically on the fruit variety, and some environmental factors like the weather, altitude, and soil. The main variety used for AO production is the Hass variety, which is also the most sold worldwide as fresh fruit. This oil can contain up to 71% of monounsaturated fatty acids (MUFA), 13% of polyunsaturated fatty acids (PUFA), and 16% of saturated fatty acids (SFA). Oleic acid is the most abundant fatty acid in the oil and thus, in the MUFA fraction (40.73 to 64.62%); the palmitic acid is the most abundant SFA (15.7%), and the linoleic acid the most abundant PUFA (13.7%) (Flores et al., 2019). As aforementioned, the main fatty acid in avocado oil is the oleic acid (between 56 to 74%), followed by linoleic acid (between 10 to 17%), palmitic acid (between 9 to 18%), palmitoleic acid (between 3 to 9%), and linolenic acid (up to 2%) (NMX-F-052-2008).

The unsaponifiable fraction of avocado oil is less than 5% of the total in the extracted oil, and it strongly depends on the extraction method used. Therefore, cold pressing and water dragging are the methods that preserve this fraction, which contains these essential secondary metabolites (Tan, 2019; Werman and Neeman, 1986). This fraction is mainly composed of glycerophospholipids, free and esterified sterols, terpenoids like squalene, carotenoids, and tocopherols, chlorophylls, and phenolic compounds.

Phospholipids or glycerophospholipids are modifications of triacylglycerols in which phosphate is found. They are minor components of the polar fraction of avocado pulp. Moreover, phospholipids are mainly composed of phosphatidylcholine, lysophosphatidylcholine, phosphatidylethanolamine, phosphatidylinositol, phosphatidic acid, and phosphatidylglycerol (Takenaga et al., 2008). This fraction is used in pharmaceutical and cosmetic preparations as emulsifiers, and it is possible to find them in no refined oils.

Phytosterols are commonly present in oils and fats as free or sterol ester of fatty acids. Avocado oil is an excellent source of this substance. According to the Mexican standard (NMX-F-052-2008), the total content of sterols is up to 4040 mg/kg of oil. The most common sterol found is β-sitosterol (between 89 to 92 mg/kg) followed by campesterol (between 6 to 8 mg/kg). Cholesterol, brasicosterol, stigmasterol, and avenasterol are also present. β-sitosterol has essential health properties since it can help to reduce plasma cholesterol concentrations by reducing the risk of cardiovascular diseases. The presence of sterols depends on the extraction method. Therefore, β-sitosterol is higher when the avocado pulp is lyophilized and cold-pressed (91.12 mg/kg).

The most common triterpenoid found in the avocado pulp and seed is squalene, which is an essential secondary metabolite with anticancer activity. Like the rest of secondary metabolites, its presence in the final oil depends on the extraction method used, and cold pressing at 40°C seems to be the best extraction condition (dos Santos et al., 2014). Tocopherols inhibit the initiation and propagation of free radicals by donating their phenolic hydrogen to lipid-free radicals. They are especially useful at high temperatures and dark conditions, they act as antioxidants during storage, and they help to maintain the shelf life of oils and can be added as preservatives (up to 0.03% of the total of the oil, in Mexico) (NMX-F-052-2008), or be naturally present in oils. Its natural presence is highly dependent on the extraction method, and cold-pressed oils are those with higher contents (dos Santos et al., 2014). However, high amounts of tocopherols added (250 ppm) act as pro-oxidants in refined avocado oil due to the effects of tocopherols. These may be overwhelmed due to photosensitized oxidation and the presence of chlorophylls and any more pro-oxidants (Werman and Neeman, 1986). Lower quantities of tocopherols added seem to have antioxidant activity. Therefore, when 100 ppm of α-tocopherol are added to the avocado purée, they delayed the peroxide formation, and the induction stage oils increased for at least 12 weeks (Elez-Martínez et al., 2005). In general, tocopherols could be found in avocado oils between 83 to 100 mg/kg and in α-tocopherols between 64 to 100 mg/kg of oil.

Moreover, the addition of tocopherols is approved up to 0.03% of the total of the oil (NMX-F-052-2008). Carotenoids also have potent antioxidant activity, and they help to reduce the incidence of various diseases. These can scavenge free radicals formed from the oxidation process and have a protective effect mainly under light conditions. Carotenoids in avocado oil are around 1.9 ug/g of product, being lutein, the primary carotenoid found (Wolf et al., 2009). Chlorophylls are extracted from the skin and the mesocarp of the fruit, these contribute significantly to the phytochemical components present in avocado oil and have shown an anticancer effect. These are firmly photosensitizers, and they promote photooxidation reactions in the oil leading the production of free radicals, they act as antioxidants during darkness storage, and are present in high quantities in avocado oils, mostly when they are cold-pressed (Wolf et al., 2009).

Polyphenols of avocado are mostly found in the seeds and the peels and are being used as added food antioxidants (Wang et al., 2019). No significant differences in the extraction of the phenolic compounds were found under different techniques, although the presence of phospholipids can interfere with the antioxidant capacity of the phenolics (Tan et al., 2019).

The physicochemical characteristics for the determination of the AO quality proposed as international standards by Wolf et al., 2009 are listed in Table 1. Up to now, the Codex Alimentarius standards for AO are the same for other fats and oils, the Codex Standard for Named vegetable oils (Codex-Stand 210-1999) is currently under review by the Codex Alimentarius Committee on fats and oils. In Mexico, there is a voluntary Mexican Standard about avocado oil, which only serves as a reference guide (NMX-F-052-2008) (Table 2).

Table 1. Physicochemical characteristics of avocado oil proposed as international quality standards

	Extra virgin	Virgin	Pure
Acid value (mg/g)	≤1	≤2.0	≤0.2
Peroxide value (mEq/kg oil)	≤4.0	<8.0	<0.5
Storage stability	Two years at room temperature when stored under nitrogen and out of the light	18 months at room temperature when stored under nitrogen and out of the light	Two years at room temperature when stored under nitrogen and out of light
Smoke point	≥250°C	≥200°C	≥250°C
Moisture	≤0.1%	≤0.1%	≤ 0.1%

Adapted from Wolf et al., 2009.

Different studies have analyzed the physicochemical properties of AO extracted by different methods such as cold pressing, Soxhlet extraction with hexane, conventional extraction with hexane, supercritical fluid extraction with CO_2 (SCO_2), and ultrasound-assisted aqueous extraction

(UAAE). A summary of the main properties analyzed by different authors is listed in Table 3.

Table 2. Physicochemical characteristics of avocado oil according to the Mexican Standards

	Avocado oil		**Pure edible avocado oil**	
	min	max	min	max
Free fatty acids (%)		1.5		0.1
Moisture (%)		0.5		0.1
Colour (Lovibond scale)		3.5		2.0 R
Density (25°C)	0.910	0.920	0.910	0.920
Peroxide index (meq/kg)		10.0		2.0
Cold test (0°C) (h)	n/a	n/a	5.5	n/a
Stability (h) OSI at 110°C	n/a	n/a	8	
Insoluble impurities (%)		0.2		0.1
Unsaponifiable matter (%)	n/a	12	1.0	1.5
Refraction index (40°C)	1.458	1.465	1.458	1.465
Iodine value	85	90	85	90
Saponification index (mg KOH/g	117	198	177	198
Mineral oil	-	-	-	-

* n/a: not applicable.

Moreover, recently, an analysis of some commercial AO sold in Mexico was carried out by the Procuraduría Federal del Consumidor (Office of the Federal Prosecutor of the Consumer) (Profeco, by its acronym in Spanish) (Profeco, 2019), finding a Peroxide Index between 4.0 to 9.8 g/ 100g and an acid value ranged from 0.04 to 1.48 mg/g. These results indicated that, currently, the AO present in the Mexican market could be considered as a virgin quality, but not as an extra virgin, due to its primary oxidation level.

The most important parameters used to define the physicochemical oxidative stability of vegetal oils during storage are those which indicate the primary oxidation:

1) Acid value or free acidity;
2) Peroxide index, and
3) Iodine value (Roncero et al., 2016).

Table 3. Physicochemical properties of avocado oil

	Li et al., 2019			Tamara de Souza et al., 1995			Tan et al., 2018		
Extraction method	Aqueous	n-hexane	Cold-pressed	Cold-pressed Margarida variety	Cold-pressed Hass variety	Commercial	n-hexane	SCO_2	UAAE
Acid value (mg/g)	0.86 ± 0.09	1.13 ± 0.07	1.16 ± 0.06	3.64 ± 0.00	1.58 ± 0.00	0.82 ± 0.00	0.38 ± 0.03	0.32 ± 0.00	0.29 ± 0.03
Iodine value (g/100g)	154.41 ± 4.45	96.90 ± 0.63	139.61 ± 1.86	80.80 ±0.00	82.87 ± 0.00	84.70 ± 0.00	75.61 ±0.91	81.70 ± 0.46	83.21 ± 0.38
Peroxide value (g/100g)	0.07 ± 0.00	0.15 ± 0.00	0.17 ± 0.00	4.98 ± 0.36	5.54 ± 0.25	4.80 ± 0.29	n/d	n/d	n/d
Moisture and volatile matter (%)	0.13 ± 0.01	15.10 ± 0.06	0.44 ± 0.11	n/d	n/d	n/d	n/d	n/d	n/d
Saponification index (mg/g)	198.0 ± 1.2	162.0 ± 9.5	187.0 ± 15.0	194.92 ± 0.02	195.79 ± 0.07	193.13 ± 0.03	186.16 ± 0.88	179.77 ± 1.05	177.80 ± 0.74
Unsaponification value (%)	0.26 ± 0.2	3.41 ± 0.24	3.66 ± 0.17	1.19 ± 0.09	0.76 ± 0.03	1.00 ± 0.08	1.58 ± 0.13	1.64 ± 0.03	1.41 ± 0.07
p-anisidine	n/d	n/d	n/d	0.04 ± 0.00	0.95 ± 0.00	0.82 ± 0.00	n/d	n/d	n/d
Oxidative stability index	n/d	n/d	n/d	3.87 ± 0.08	5.57 ± 0.07	6.31 ± 0.18	n/d	n/d	n/d

*n/d: not determined.

Adapted from different authors.

The acid value (AV) is the quantity of potassium hydroxide (mg) necessary to neutralize the fatty acids in 1 g of sample, and it is determined by titration of the oil in ethanol and generally expressed in oleic acid (Endo, 2018). Another method to determine the oil´s acidity is the UV absorbance test, mainly used to determine the age of refinement of the oils. This test measure changes between 232 and 270 nm caused by peroxides (Wiesman et al., 2009). The value proposed for extra virgin AO is slightly higher than for extra virgin olive oil (0.8 g/100g). Most of the oils tested overcame the limit suggested for extra virgin AO indicating failures during processing and storage.

The peroxide index (PI) gives initial evidence of the rancidity of the unsaturated fatty acids, mainly in the first phase of oxidation, generally known as the induction period. Hydroperoxides reacted with iodide ions to form iodine, and the PI is determined by titration of the liberated iodine with thiosulphate by using starch as an indicator. The rate of oxidation depends on the degree of saturation of the fatty acids as well as the number of antioxidants present. The value proposed for AO is lower than for virgin olive oil (<20 meq/kg). However, it is considered that a PI higher than ten meq/kg is indicative of low, stable oil.

The iodine value (IV) is used to measure the degree of unsaturation of oils and fats. Results are expressed as the number of grams of iodine absorbed by 100 g of oil. The IV indicates the degree of unsaturation of the oil, and therefore, a higher value is associated with a more significant potential of oxidation. The value decreases with the temperature and moisture during the storage (Roncero et al., 2016).

Other physicochemical determinations include the measurement in the formation of secondary oxidation products, which in general are more stable than the primary ones. This determination can include the thiobarbituric acid method and the p-anisidine value, both by UV-vis spectroscopy and the specific determination of secondary products of interest by chromatography (Barriuso et al., 2013).

There are new analytical approaches, including the use of chemometric tools. RAMAN and IR spectroscopy and Magnetic Nuclear Resonance are being explored better to determine the physicochemical characteristics of oils and fats, and hence, their final quality (Alvarenga et al., 2018).

3. Oxidative Stability of the AO

Rancidity is the leading cause of deterioration of edible oils and fats with the production of unpleasant tastes and smells, changes in color, viscosity, density, and solubility. Further consequences of oil oxidation include nutritional losses like the degradation of essential fatty acids, vitamins, and pro-vitamins (Matthaus, 2010). The main reactions resulting in oxidative deterioration of oils are:

1) Auto-oxidation;
2) Photooxidation, stimulated by the presence of chlorophylls;
3) Enzymatic oxidation, and
4) Irradiation.

These four reactions are commonly produced during the extraction and storage in adverse conditions of edible oils, meaning primarily high temperatures, high humidity, presence of oxygen and light (Pathirana et al., 2013). Internal factors such as the degree and nature of the unsaturation of its glycerides, the antioxidant content, and the presence of pro-oxidants can also affect the oxidation process (Kochhar and Henry, 2009).

The auto-oxidation is the oxidative deterioration of edible oils and fats, which occurs slowly at room temperature. It is a complex mechanism prevalent in the degradation of unsaturated fats ruled by the presence of oxygen, which produces hydroperoxides and cyclic organic peroxides. The oxidation is initiated by the formation of radicals as a result of the homolytic splitting of hydrogen atoms in the alpha position concerning the double bond. The extra virgin avocado oil extracted by cold pressing is very susceptible to the photooxidation process due to their high amount of chlorophylls and relative low quantity of antioxidants like polyphenols, carotenoids, and tocopherols. Therefore, avocado oil is susceptible to light oxidation in contrast with its high stability in the dark at room temperature, mainly because the chlorophylls content in the avocado is drastically reduced with the light exposition (Werman and Neeman, 1986). Most of the proteins present in the avocado are oxidative enzymes as lipoxygenase, lipase, and polyphenol oxidase, which are present in the pulp and produce degradation when it is exposed to the oxygen (Vargas-Ortiz et al., 2017). Lipoxygenases act by degrading the essential omega-three fatty acids like linolenic acid more than the essential linoleic acid (an omega-six fatty acid), which are essential structural components of cell membranes and serve as precursors to bioactive lipid mediators. The optimum activity of the enzyme is given at 40°C and pH 6.5, which is a typical extraction temperature in high-quality cold-pressed oils (Jacobo-Velázquez et al., 2010). Finally, ionizing radiation can cause oxidation of polyunsaturated fatty acids (especially in omega 3) (Hammer and Willis, 1979), although it is commonly used to extend the shelf life of the fruit in post-harvest.

The presence of light has a strong influence on the rancidity of AO since the high levels of chlorophylls present act as photosensitizers promoting oxidation, while the presence of carotenoids and tocopherols can inhibit the oxidation process. The exposure to light stimulates the chlorophylls, which react with lipids forming intermediates that can form free radicals and oxidation-breakdown products (Wolf et al., 2009). The presence of oxygen also has a negative influence on the stability because it can catalyze the autoxidation and the photooxidation. Moreover, cooking at high temperatures can promote the oxidation and degradation of unsaturated fatty acids, phytosterols, and antioxidants. The conditions of deterioration depend on the culinary process, which includes temperature, time, and food processes (Berasategui et al., 2012). Consequently, cooking processes, which include the presence of high amounts of water, thus having a more potent oxidant activity.

The main products of lipid oxidation are summarized in Table 4.

Table 4. The main products of lipid oxidation present in food products

Primary products of lipid oxidation	Secondary products of lipid oxidation
Non-esterified hydroperoxyl fatty acids like hydroperoxyl linoleate 13-HPODE	alcohols
Non-esterified hydroxy fatty acids like hydroxy linoleate-13-HODE	aldehydes
Non-esterified epoxy fatty acids like 9,10-epoxy linoleate	ketones, hydrocarbons
Oxidized triacylglycerol (TAG) monomers and dimers	hydrocarbons
Cyclic TAG monomers and dimers	

Adapted from Jackson and Penumetcha, 2019.

The oxidized PUFA are related to different health problems at the cell and tissue level. Oxidized PUFA can be consumed, which increases the amount of malondialdehyde in urine as well as of lipophilic carbonyl compounds (Kanner, 2007), or can be formed in the gastrointestinal tract, especially in unhealthy individuals (Vieira et al., 2017).

The effect of oxidized PUFA coming from the diet has been scarcely studied in humans. However, studies in animal models suggest that a diet containing thermally oxidized oils increases the risk of oxidative stress, atherosclerosis, insulin resistance, glucose intolerance, and hypertension (Jackson and Penumetcha, 2018).

4. Shelf Life of Avocado Oil

According to the Institute of Food Technologists (IFT), the shelf life of a food product is defined as the period between the manufacture and the retail purchase of a food product, during which time the product has a state of satisfactory quality in terms of nutritional value, taste, texture, and appearance. In oils, it is defined as the length of time under normal storage conditions with no off-flavors or defects developed, and quality parameters are within the accepted limits for its commercial category (Li and Wang 2018). The shelf life is controlled by three factors:

1) Intrinsic parameters like the composition of the product;
2) Extrinsic parameters such as the environment to which the product is exposed to, like temperature, air, light, etc., and
3) Package properties.

All oils and fats oxidize during the storage. The rate of oxidation depends on inherent factors, such as the fatty acid profile, the presence of minor components including moisture and metals, and external factors like the storage temperature, the air exposition, and the sunlight.

As aforementioned, the lipid oxidation reduces the shelf life of foods. Thus unpleasant odors and tastes are produced and can limit the nutritive value by limiting the content of essential PUFA (Ahmed et al., 2016).

The oxidative stability or rancidity is the most critical parameter to determine the shelf life of edible oils and fats. The initial phases of lipid oxidation can be detected by measuring the peroxide value (PV), which quantifies the formation of peroxides and hydroperoxides. The formation of secondary oxidation products can be determined through the anisidine value, the carbonyl value, and the thiobarbituric (TBARS) acid value. All these methods are sensitive and straightforward, reliable, and reproducible (Maszewska et al., 2018). Other methods can also estimate the oxidation as the determination of the acid value, the epoxide or oxirane value, and the iodine value. Recently, alternative methodologies have been explored to quantify and characterize the lipid oxidation products. Between them, different spectroscopic methods are probably the most successfully used due to the minimal sample treatment required. Therefore, chemiluminescence, fluorescence, infrared, and Raman spectroscopic associated methods are widely used (Xu et al., 2015). Nuclear magnetic resonance and electron paramagnetic resonance, as well as chromatographic methods (HPLC, CG-MS, among others), have been used to identify and quantify individual products of oxidation, which allows the understanding of oxidation reactions (Barriuso et al., 2013).

The determination of shelf life can be very time-consuming when analyzed at room temperatures since the oil shelf life can be up to 18 months. Thus, the use of accelerated techniques is an increasingly used option to determine the best-before date. Several parameters can be used to measure the oil's rancidity at high temperatures (60-140°C), presence of oxygen through a pressure increase (3-165 psi), and presence of metal traces (5-10 ppm). Several accelerated tests have been described to assess the oxidative stability of edible oils, such as the Schaal oven test, Active oxygen method, and Rancimat. Between them, the Rancimat is the most used for its simplicity and reproducibility (Farhoosh, 2007).

To determine the shelf life of the oil, two prediction models are used to simulate oil degradation: kinetic models and empirical models. Kinetic models analyze the effect of some stressing conditions (high temperature, light, and oxygen) on critical chemical parameters (free fatty acids, peroxide value, ultraviolet absorbance, induction time, phenol degradation, among others). The empirical models study the correlation between individual chemical parameters and experimental conditions, by using advanced statistical analysis to build regression models that enable the prediction of maximum shelf life as a function of chemical parameters (Li et al., 2019).

The shelf life of a culinary avocado oil purchased from a UK supermarket has been established by producers in 13 months. However, Kochhar et al., 2009 used the lineal Arrhenius law to extrapolate the oil degradation with a Pearson correlation coefficient (R2) of 0.9991 in seven months, which is significantly lower than the before-date given by producers. Moreover, using the Rancimat test, they found that the induction period at 100°C was 16.9 h and the Q10 factor, which is the decrease in shelf life for each 10°C of increase in temperature, of 2.05. Recently, similar results were found by Aktar and Adal, 2019, when they determined the Arrhenius kinetics of avocado oil extracted by cold pressing (60°C, 45 min), also by using the Rancimat test conditions. They found a predicted shelf life of 210 days (7 months) at 25°C and a Person correlation coefficient (R2) of 0.9991. The induction time was calculated at five different temperatures, and 100°C was 15.02 h slightly inferior to those reported by Kochhar et al., 2009. According to these studies, the avocado oil follows the Arrhenius relationship over a temperature range of 100 to 140°C.

To increase the shelf life of avocado oil, the addition of natural antioxidant substances has been explored. This is the case of the addition of caffeic acid and p-coumaric acid (Sun-Waterhouse et al., 2011). These two phenolic compounds have a relatively high antioxidant capacity and may suppress the lipid oxidation, by the donation of hydrogen atoms to the lipid peroxyl radicals, to interfere with the initiation or propagation of primary oxidation. The results showed that the addition of these phenolic compounds have little effect on avocado oil at 20°C, but showed a protective effect at 60°C. Then, the addition of caffeic or coumaric acid could maintain the nutritional value of avocado oil during storage.

Conclusion

The shelf life of avocado oil is ruled basically by the lipidic oxidation, which depends on the presence of secondary metabolites in the oil and the storage conditions. The presence of secondary metabolites, meaning the unsaponifiable fraction is related to the extraction method used. Therefore, in general, cold-pressed oils have a higher amount of biometabolites since the stressful conditions during Soxhlet extraction oxidize some of them, and the refining process eliminates others from the final oil. Moreover, the high presence of chlorophylls in avocado oil, mostly in virgin and extra virgin oils, make it a photosensitizer. Hence, the oil must be stored in the dark to prolong its shelf life. High temperatures and the presence of oxygen can also improve the rancidity of the oil, so it should be stored at room temperature (20-25°C) or at least in static conditions to reduce the oxygen reactivity. The addition of natural antioxidants (tocopherols, citric acid, and carotenoids) has been explored with different results. Therefore, the use of 100 ppm of tocopherols acted as antioxidants, but a higher concentration of 250 ppm acted as pro-oxidants.

Despite its growing commercial importance, there are few types of research about the lipid oxidation of avocado oil during storage and cooking. The few mathematical models to predict its shelf life are based in the lineal Arrhenius law and have determined an optimal storage life of the virgin oil of seven months, which is significantly lower than the 15 months usually given by the producers in the commercial oil. In this regard, it is necessary to have a better knowledge of the structure and activity of the oxidative stability of avocado oil. Studies about how to reduce the presence of chlorophylls in avocado oil, especially in cold-pressed, should also be explored.

Finally, avocado oil has excellent economic potential, due to its bio-functionality as a preventive agent of some cardiovascular and metabolic diseases and for its great techno-functionality, which includes its high thermal stability during cooking and its neutral flavor. Hence, more studies focusing on some of its technological aspects should be carried out.

References

Ahmed, M., Pickova, J., Ahmad, T., Liaquat, M., Farid, A., & Jahangir, M. (2016). Oxidation of Lipids in Foods. *Sarhad Journal of Agriculture,* 32(3).

Aktar, T., and Adal, E. (2019). Determining the Arrhenius Kinetics of Avocado Oil: Oxidative Stability under Rancimat Test Conditions. *Foods,* 8(7), 236. https://doi.org/10.3390/foods8070236.

Alvarenga, B. R., Xavier, F. A. N., Soares, F. L. F., and Carneiro, R. L. (2018). Thermal Stability Assessment of Vegetable Oils by Raman Spectroscopy and Chemometrics. *Food Analytical Methods,* 11(7), 1969–1976. https://doi.org/10.1007/s12161-018-1160-y.

AOCS, Official Methods and Recommended Practices of the American Oil Chemists' Society. (2004). American Oil Chemists' Society, Champaign, IL.

Barriuso, B., Astiasarán, I., and Ansorena, D. (2013). A review of analytical methods measuring lipid oxidation status in foods: a challenging task. *European Food Research and Technology,* 236(1), 1–15. https://doi.org/10.1007/s00217-012-1866-9.

Berasategi, I., Barriuso, B., Ansorena, D., and Astiasarán, I. (2012). Stability of avocado oil during heating: Comparative study to olive oil.

Food Chemistry, 132(1), 439–446. https://doi.org/https://doi.org/10.1016/j.foodchem.2011.11.018.

Bollinedi, H., Singh, A. K., Singh, N., S, G. K., Bhowmick, P. K., K. K, V., … R. K, E. (2020). Genetic and genomic approaches to address rapid rancidity of rice bran. *Critical Reviews in Food Science and Nutrition,* 1–10. https://doi.org/10.1080/10408398.2020.1718598.

Campanella, L., Favero, G., Pastorino, M., and Tomassetti, M. (1999). Monitoring the rancidification process in olive oils using a biosensor operating in organic solvents1This paper was presented at the Fifth World Congress on Biosensors, Berlin, Germany, 3–5 June 1998.1. *Biosensors and Bioelectronics,* 14(2), 179–186. https://doi.org/https://doi.org/10.1016/S0956-5663(98)00117-1.

Cerecedo-Cruz, L., Azuara-Nieto, E., Hernández-Álvarez, A., González-González, C., and Melgar-Lalanne, G. (2018). Evaluation of the oxidative stability of Chipotle chili (Capsicum annuum L.) oleoresins in avocado oil. *Grasas y Aceites,* 69(1), e240. http://dx.doi.org/10.3989/gya.0884171.

Codex Stan 210-1999 (Revision 2009. Amendment, 2013). Codex Standard for named vegetable oils.

Dos Santos, M. A. Z., Alicieo, T. V. R., Pereira, C. M. P., Ramis-Ramos, G., and Mendonça, C. R. B. (2014). Profile of Bioactive Compounds in Avocado Pulp Oil: Influence of the Drying Processes and Extraction Methods. *Journal of the American Oil Chemists' Society,* 91(1), 19–27. https://doi.org/10.1007/s11746-013-2289-x.

Elez-Martinez, P., Soliva-Fortuny, R. C., Gorinstein, S., and Martin-Belloso, O. (2005). Natural Antioxidants Preserve the Lipid Oxidative Stability of Minimally Processed Avocado Purée. *Journal of Food Science,* 70(5), S325–S329. https://doi.org/10.1111/j.1365-2621.2005.tb09986.x.

Endo, Y. (2018). Analytical Methods to Evaluate the Quality of Edible Fats and Oils: The JOCS Standard Methods for Analysis of Fats, Oils, and Related Materials (2013) and Advanced Methods. *Journal of Oleo Science,* 67(1), 1–10. https://doi.org/10.5650/jos.ess17130.

Farhoosh, R. (2007). Shelf-life prediction of edible fats and oils using Rancimat. *Lipid Technology,* 19(10), 232–234. https://doi.org/10.1002/200700073.

Flores, M., Saravia, C., Vergara, C., Avila, F., Valdés, H., and Ortiz-Viedma, J. (2019). Avocado Oil: Characteristics, Properties, and Applications. *Molecules,* 24(11), 2172. https://doi:10.3390/24112172.

Hammer, C. T., and Wills, E. D. (1979). The Effect of Ionizing Radiation on the Fatty Acid Composition of Natural Fats and on Lipid Peroxide Formation. *International Journal of Radiation Biology and Related Studies in Physics, Chemistry and Medicine,* 35(4), 323–332. https://doi.org/10.1080/09553007914550391.

Jackson, V., and Penumetcha, M. (2019). Dietary oxidized lipids, health consequences, and novel food technologies that thwart food lipid oxidation: an update. *International Journal of Food Science and Technology,* 54(6), 1981–1988. https://doi.org/10.1111/ijfs.14058.

Jacobo-Velázquez, D. A., Hernández-Brenes, C., Cisneros-Zevallos, L., and Benavides, J. (2010). Partial purification and enzymatic characterization of avocado (*Persea americana* Mill, cv. Hass) lipoxygenase. *Food Research International,* 43(4), 1079–1085. https://doi.org/https://doi.org/10.1016/j.foodres.2010.01.021.

Jorge, Tamara de Souza, Polachini, Tiago Carregari, Dias, Liara Silva, Jorge, Neuza, and. Telis-Romero, Javier. (2015). *Physicochemical and rheological characterization of avocado oils. Ciência e Agrotecnologia,* 39(4), 390-400. https://doi.org/10.1590/S1413-70542015000400010.

Kanner, J. (2007). Dietary advanced lipid oxidation endproducts are risk factors to human health. *Molecular Nutrition and Food Research,* 51(9), 1094–1101. https://doi.org/10.1002/mnfr.200600303.

Kochhar, S. P., and Henry, C. J. K. (2009). Oxidative stability and shelf-life evaluation of selected culinary oils. *International Journal of Food Sciences and Nutrition,* 60(sup7), 289–296. https://doi.org/10.1080/09637480903103774.

Krumreich, F. D., Borges, C. D., Mendonça, C. R. B., Jansen-Alves, C., and Zambiazi, R. C. (2018). Bioactive compounds and quality parameters of avocado oil obtained by different processes. *Food Chemistry,* 257, 376–381. https://doi.org/https://doi.org/10.1016/j.foodchem.2018.03.048.

Li, X., & Wang, S. C. (2018). Shelf Life of Extra Virgin Olive Oil and Its Prediction Models. *Journal of Food Quality,* 2018, 1639260. https://doi.org/10.1155/2018/1639260.

Li, X., and Wang, S. C. (2018). Shelf Life of Extra Virgin Olive Oil and Its Prediction Models. *Journal of Food Quality,* 2018, 1639260. https://doi.org/10.1155/2018/1639260.

Li, Y., Liu, Y., Deng, D., Liang, J., Chen, W., Chen, X., & Li, J. (2019). Study on Extracting Avocado Oil from Avocado Pulp by Aqueous Extraction. *IOP Conference Series: Earth and Environmental Science,* 330, 42027. https://doi.org/10.1088/1755-1315/330/4/042027.

Maszewska, M., Florowska, A., Dłużewska, E., Wroniak, M., Marciniak-Lukasiak, K., & Żbikowska, A. (2018). *Oxidative Stability of Selected Edible Oils. Molecules,* 23(7), 1746. doi:10.3390/molecules23071746.

Matthäus, B. (2010). 6 - Oxidation of edible oils. In E. A. Decker, R. J. Elias, & D. B. T. O. in F. and B. and A. A. Julian McClements (Eds.), *Woodhead Publishing Series in Food Science, Technology and Nutrition* (pp. 183–238). https://doi.org/https://doi.org/10.1533/9780857090331.2.183.

Matthäus, B. (2010). 6 - Oxidation of edible oils. In E. A. Decker, R. J. Elias, & D. B. T. O. in F. and B. and A. A. Julian McClements (Eds.), *Woodhead Publishing Series in Food Science, Technology and Nutrition* (pp. 183–238). https://doi.org/https://doi.org/10.1533/9780857090331.2.183.

NMX-F-052-SCFI-2008. Aceites y grasas. Aceite de aguacate-especificaciones. *Fats and oils.* Avocado oil-specifications.

Piccinin, E., Cariello, M., De Santis, S., Ducheix, S., Sabbà, C., Ntambi, J. M., and Moschetta, A. (2019). Role of Oleic Acid in the Gut-Liver Axis: From diet to the Regulation of Its Synthesis via Stearoyl-CoA Desaturase 1 (SCD1). *Nutrients,* 11(10), 2283. https://doi:10.3390/11102283.

Profeco. (2019). Aceites comestibles. *Revista del consumidor.* 9, 4-20. (retrieved from https://issuu.com/profeco/docs/revista_del_consumidor_septiembre_2019, April 04,2020).

Roncero, J. M., Álvarez-Ortí, M., Pardo-Giménez, A., Gómez, R., Rabadán, A., and Pardo, J. E. (2016). Almond virgin oil: Parameters of regulated physicochemical quality and stability. *Rivista Italiana delle Sostanze Grasse,* 93: 237-243.

Sun-Waterhouse, D., Thakorlal, J., and Zhou, J. (2011). Effects of added phenolics on the storage stability of avocado and coconut oils. *International Journal of Food Science and Technology,* 46(8), 1575–1585. https://doi.org/10.1111/j.1365-2621.2011.02655.x.

Takenaga, F., Matsuyama, K., Abe, S., Torii, Y., and Itoh, S. (2008). Lipid and fatty acid composition of mesocarp and seed of avocado fruits harvested at northern range in Japan. *Journal of Oleo Science,* 57(11), 591–597. https://doi.org/10.5650/jos.57.591.

Tan, C. X. (2019). Virgin avocado oil: An emerging source of functional fruit oil. *Journal of Functional Foods*, 54, 381–392. https://doi.org/https://doi.org/10.1016/j.jff.2018.12.031.

Unlu, N. Z., Bohn, T., Clinton, S. K., and Schwartz, S. J. (2005). Carotenoid Absorption from Salad and Salsa by Humans Is Enhanced by the Addition of Avocado or Avocado Oil. *The Journal of Nutrition,* 135(3), 431–436. https://doi.org/10.1093/jn/135.3.431.

Vargas-Ortiz, M., Servent, A., Salgado-Cervantes, M., and Pallet, D. (2017). Stability of the lipid fraction of avocado puree obtained by flash vacuum-expansion process. *Innovative Food Science and Emerging Technologies.*

Vieira, S. A., Zhang, G., and Decker, E. A. (2017). Biological Implications of Lipid Oxidation Products. *Journal of the American Oil Chemists' Society,* 94(3), 339–351. https://doi.org/10.1007/s11746-017-2958-2.

Vingering, N., Oseredczuk, M., du Chaffaut, L., Ireland, J., and Ledoux, M. (2010). Fatty acid composition of commercial vegetable oils from the French market analyzed using a long highly polar column. *OCL,* 17(3), 185–192. https://doi.org/10.1051/ocl.2010.0309.

Visioli, F., Davalos, A., López de las Hazas, M. C., Crespo, M. C., and Tomé-Carneiro, J. (2020). An overview of the pharmacology of olive oil and its active ingredients. *British Journal of Pharmacology,* 177(6), 1316–1330. https://doi.org/10.1111/bph.14782.

Werman, M. J., & Neeman, I. (1986). Oxidative stability of avocado oil. *Journal of the American Oil Chemists' Society,* 63(3), 355–360. https://doi.org/10.1007/BF02546046.

Wiesman, Z. (2009). *Desert olive oil cultivation: advanced biotechnologies*. Academic Press.

Wong, M., Ashton, O. B. O., McGhie, T. K., Requejo-Jackman, C., Wang, Y., and Woolf, A. B. (2011). Influence of Proportion of Skin Present During Malaxing on Pigment Composition of Cold Pressed Avocado Oil. *Journal of the American Oil Chemists' Society,* 88(9), 1373–1378. https://doi.org/10.1007/s11746-011-1790-3.

Woolf, A., Wong, M., Eyres, L., McGhie, T., Lund, C., Olsson, S., … Requejo-Jackman, C. (2009). *2 - Avocado Oil* (R. A. Moreau & A. B. T. G. and H. P. S. O. Kamal-Eldin, eds.). https://doi.org/https://doi.org/10.1016/B978-1-893997-97-4.50008-5.

In: Avocado: Consumption and Health ISBN: 978-1-53618-204-0
Editors: M. G. Melgar Lalanne et al.

Chapter 5

BIOTECHNOLOGY AND *IN VITRO* CULTURE FOR THE SUSTAINABILITY OF AVOCADO AGROECOSYSTEMS (*PERSEA AMERICANA* MILL)

María de Jesús Martínez Hernández*
Facultad de Ciencias Agrícolas, Universidad Veracruzana,
Xalapa, Veracruz, México

ABSTRACT

The social economy is based upon various services that natural resources provide at no cost. Natural resources constitute the planet's life support system, and it is precisely their accelerated degradation, which leads to major environmental problems. It would not be exaggerated to affirm that our future depends on the ability to defend and use biodiversity sustainably. Nonetheless, most people have not yet understood the role of nature, and human influence subdues the environment more and more. It becomes then necessary to consider

* Corresponding Author's Email: mhernandezmj@gmail.com.

methodologies and techniques that allow the preservation of agricultural genetic resources. Cultivated diversity—or agro diversity— is the part of biodiversity that nourishes man and which in turn is nourished by man. The *Lauraceae* family stands out in this enormous vegetable diversity, being the avocado its main product. The sale of fresh fruit and its byproducts generates significant revenue for the producer country. However, this agroecosystem is attacked by different diseases and plagues. Due to the preceding, the preservation and care of the avocado agroecosystem are essential. Consequently, techniques derived from plant biotechnology have been put into practice: in addition to providing an endless number of pathogen-free plants, they also offer the possibility of carrying out multiple studies, from ecological to molecular. Moreover, they make it possible to create *in vitro* germplasm banks that can be short-term or long-term storage for the preservation of the three avocado horticultural races.

Keywords: biotechnology, avocado, preservation techniques

1. INTRODUCTION

The origins of biotechnology can be traced back to the dawn of human history. During the Stone Age, our primitive ancestors began using live organisms and their products. Biotechnology is a term given to the evolution and the recent progress in genetic science, which originated towards the end of the 20th Century with the works by Gregor Johan Mendel. However, history began with Charles Darwin, considered to be the father of modern biology. Darwin concluded that species are not fixed and unchangeable but capable of evolving to produce new species. According to his observations, this evolution is explained because the members of a particular species displayed significant variations amongst themselves. Some of these members were, unlike others, more fit to the environment in which they were found. The fittest would produce more descendants than the rest. This process known as natural selection implied the modification of the population's characteristics so that the strongest traits would be maintained and propagated, whereas the weakest would become less frequent and would eventually disappear.

The monk Gregor J. Mendel (1822-1884) worked in his monastery's garden in Austria without being aware of the importance of his work. He chose a common plant, peas (*Pisum sativum*), as the subject of study. In his studies, instead of analyzing the overall transmission of the plant's characteristics, he paid attention to only one feature at a time, which allowed him to select certain aspects of the plant that displayed distinguishable alternatives such as the seed's shape (rough/smooth) or color (yellow/green) (Corcos, 1995).

In 1866, Mendel published the results of his experiments carried out during seven years in the garden of his Augustinians' monastery. This allowed Mendel to overcome the previous conceptions of inheritance. The primary conception then was that the characteristics were transmitted from parents to children through a series of blood-related fluids. So, when the blood was mixed in the descendants, the characters of the parents merged and could not be separated again (Guarino, 1995; Howell, 1998).

Mendel presented a new conception of inheritance. This conception proposed the idea that characteristics are not inherited as such and that only the factors that determine them are transmitted. His study on the behavior of hereditary factors was being carried out, with total intuition, 50 years before knowing about the nature of these factors (later called genes). Although he described the essential behavior of genes, his experiments did not reveal the chemical nature of the units of heredity, which occurred about half a century later, around the middle of the 20th Century, and it involved many works by different scientists around the world for several decades.

2. Impact of Biotechnology on Agriculture

Ten thousand years ago, empirical knowledge (seeding) allowed the birth of agriculture. In the 18th Century, a scientific discovery (plants have biological sex) came to the aid of breeders to obtain the varieties needed by the new agriculture that was being founded by that time (Gómez, 1999; Gonzáles, 1999). The birth of each agriculture has been accompanied by a new breeding technique (mass selection at the beginning, cross-breeding in the 18th Century) that has made it possible to "leap" to produce a new pattern of varieties. At present, the same is happening again.

New agriculture requiring a new breeding method to add to the others, not to replace them, is needed. This method exists since the beginning of the '70s and was developed regardless of practical needs (the same thing happened to the discovery of sexuality in plants). This is what is globally known as Biotechnology, although expressions such as genetic engineering, recombinant DNA, etc. are frequently employed.

With modern biotechnology, genetic manipulation ranges from the cultivation of protoplasts (cells without cell walls), cells and tissues in vitro, to the direct manipulation of DNA, which includes different techniques such as the cultivation and fusion of protoplasts, micropropagation, in vitro selection against adverse biotic or abiotic agents, and gene insertion by genetic transformation (Pérez, 1998). The implementation of these techniques was carried out in the early '70s, and it has represented an authentic breakthrough for the roof of possibilities offered by traditional biology. It is possible to transfer a single gene, even parts of the gene such as, for example, the promoter necessary to "turn on" the gene and put it to work. Moreover, this transfer is made regardless of the donor and recipient organisms.

For example, the insulin that exists today is human; it is no longer a pig's insulin as it was years ago since they succeeded in inserting the insulin production human gene in a bacterium's chromosome, which is the one that industrially produces insulin. The first human insulin appeared on the market in 1982, and, since then, several industrial products and other drugs (e.g., human growth hormone, so popular, yet illegal, amongst sportspeople) have been created.

Kjellsson et al., (1997) mention that in this form, it has been possible to obtain transgenic varieties that contain a single gene from any other organism, regardless of the possibility of frequent sexual interbreeding. It is said that such varieties will eliminate the remaining diversity. The weak transgenic plants have only one gene that differs from the varieties they derive from. In the environments where these unique genes are needed, of course, they will be eliminated, but due to their economic and environmental interest. If the reader grows corn and faces the corn borer problem, or cotton with the worms in the capsule, which are difficult to eliminate even with aggressive pesticides, what will they do if they are offered varieties resistant to these insects? They will save on pesticides, purify the environment, and improve their profit. In an endemic region for these problems, these pesticides will be employed whether or not they have a sweep away effect on other genes since the breeder will think, in a reasonable way, that if genetic diversity is a public good, the problem of its loss is also a public problem that corresponds to public authorities.

Plant's biotechnology, both in the research and economic activity arenas, contributes efficiently in the increase of agricultural production, thus making the improvement of crops possible in terms of health and quality, and reducing the employment of conventional agricultural practices that consume energy and pollute the environment (Shemaefsky, 2013). Similarly, Narváez (2015) affirms that biotechnology must provide sustainability with resource preservation practices in favor of the wellbeing of all living creatures.

Modern agricultural biotechnology comprises a variety of tools used by scientists to understand and manipulate the genetic structure of organisms that shall be employed in the production or elaboration of agricultural products. Thus, we can observe that biotechnology can be classified in five broad areas: Human Health Biotechnology (where Food B. is included), Animal Biotechnology, Industrial Biotechnology, Plant Biotechnology, and Environmental Biotechnology. On the other hand, DaSilva (2004) refers to a classification of biotechnology through colors: green concentrates studies in agriculture, environmental biotechnology, biofuels, biofertilizers, bioremediation, and geomicrobiology.

3. Benefits of Biotechnology

Plant biotechnology offers the possibility of producing crops that will not only taste better but will also be healthier. On the other hand, Marx

(1989) mentions that agronomic traits incorporated by biotechnology (such as insect or herbicide resistance) increase the agricultural value of crops by allowing them to do things that increase production or reduce the need for other inputs such as agrochemicals or fertilizers. Among our products that currently have built-in features that provide higher yields at lower costs (since they allow better insect and weed control), we are already cultivating potatoes that require 40% less chemical insecticides than conventional varieties (Kelly & George, 1998).

Quality or "profit" attributes help value the product for consumers by improving the quality of the food and the fiber produced by the plant. It is likely that in the future, it will be possible to offer potatoes that absorb less oil when fried, corn, and soybeans containing a more significant amount of proteins, tomatoes with a fresher flavor, and strawberries that retain their natural sweetness. Someday, seeds will become unparalleled, energy-efficient, and environment-friendly production centers capable of generating products that at present come from non-renewable resources. A spring rapeseed (canola) plant, for example, may become a factory that adds beta-carotene to canola oil to alleviate the nutritional deficiency that causes night blindness.

Kelly and George (1998) claim that biotechnology innovations will triple crops' yield without requiring additional farmland, thus saving natural forests and animal habitats. Other innovations can reduce or eliminate the dependence on agrochemicals that may contribute to environmental degradation—others will preserve soil and water resources (Mohan et al., 1998). Many experts agree that the world cannot afford to wait any longer before taking action. If we act now by developing the technology and infrastructure needed to meet the future needs of humankind, we can feed the world for the centuries to come and improve the quality of life for people around the world. The positive visualization of the future of the sustainable application of agricultural biotechnology is framed within the global trends such as healthy food, 100% natural products with good appearance and high nutritional quality, as well as the reduction of the employment of additives and chemical ingredients, fresh products, and long-lasting or minimally processed products, a more ecological agrochemical-free agriculture and functional food, fortified and enriched with high fiber content, dietary, and energy. (Gómez et al., 2013)

4. FUTURE OF BIOTECHNOLOGY

In the 21st Century, agriculture is facing unprecedented challenges. Over the next 30 years, an additional two billion people will have to be fed from an increasingly fragile natural resource base. Over than 842 million people, most of whom live in rural areas of developing countries, are chronically deprived of food, and many millions more are malnourished due to the poor quality of their regular diet or lack of diversity. However, genetic engineering is proclaimed as a critical tool for addressing food insecurity and malnutrition in developing countries, and their adversaries are accused of crimes against humankind for delaying the regulatory approval of potentially life-saving innovations (Potrykus, 2003).

Biotechnology makes it possible to overcome the most problematic or difficult production constraints with breeding methods. Biotechnology allows to accelerate conventional breeding programs by providing farmers with disease-free planting material, by creating crops that are resistant to pests and diseases in situations where chemicals are harmful to the environment and human health, by providing diagnostic tools and vaccines that help combat devastating animal diseases; it improves the nutritional quality of basic food as well as the creation of new products for health and industrial purposes (Dreher et al., 2000; Freeze, 2002). It is also used in the livestock and fishery sectors.

5. METHODOLOGICAL APPROACHES TO ASSESS AGROECOLOGICAL SUSTAINABILITY

The word sustainability has different meanings for different people; however, there is consensus that it has an ecological basis. In a general way, sustainability is a version of the concept of "sustained yield," that is, the condition or capacity to harvest in perpetuity specific biomass from a system that can renew itself or which renewal is not at risk (Gliessman, 1990; Bosshard, 2000). In the broad sense, agroecological sustainability can be considered as that sustainability that integrates ecological, social, and economic aspects that include directly or indirectly in agriculture; in other words, it considers livestock, forestry, aquaculture, and agriculture *per se*. Considering that, at present, the ecological aspect of agroecology must integrate *Homo sapiens* as another organism of the ecosystem—since it is the one that participates as a decision-maker and/or controller of the agroecosystem and not as an alien or external element. Nonetheless, in general terms, it can be said that even sustainability measures are partial, that they focus on taking into account the ecological, social, and economic aspects in an isolated manner. However, some contributions already try to integrate these factors into a single scheme of analysis and evaluation. On the other hand, it is necessary to take into account that the analysis degree and depth shall depend on the time, human and economic resources available for this purpose.

Assessing agricultural sustainability has not been an easy task, as shall be discussed below. The measures and points of view on agricultural sustainability vary; they depend on the researchers' education and the interest of institutions, ranging from global to local scales according to how researchers consider the complexity of interactions that influence agricultural and non-agricultural activities. There are several suggestions for assessing agricultural sustainability as a result of unidisciplinary, multidisciplinary, and inter-disciplinary work. Some of these combine philosophical, ethical, and empirical approaches (Gliessman, 1990; Bosshard, 2000; Ruiz, 2001).

Sustainable agriculture has been evaluated by different approaches. There are useful reports that tend to analyze sustainability from a disciplinary point of view, as indicated by Lee (1994), where they individually evaluate water, soil, insect pests, and financial aspects. Campbell et al. (1995) also suggest a variety of biophysical and desirable measures to assess the sustainability of temperate zone agroecosystems.

Long-term experiments and research have focused on economic and environmental sustainability, statistically demonstrating how high yields can be maintained over time. Hence, modern agriculture feeds six billion people. Cereal production has doubled in the last 40 years, mainly due to the increase in yields achieved from the so-called "green revolution," which consisted of the use of more significant amounts of fertilizers, irrigation, fertilizer-tolerant varieties, herbicides, insecticides, fungicides and other technologies applied to agriculture.

The population projected for 2050 is 50% higher than today. The global demand for grains shall double due to an increase in people's income and changes in the diet towards meat (which will require grain-based nutrition). Meeting these demands with out affecting the remaining forest ecosystems while maintaining the sustainability of agroecosystems is a challenge for humanity.

Given this future scenario, biotechnology will undoubtedly be an essential element to fulfill the growing demand for quantity and quality of food and the sustainability of agroecosystems in all their multiple applications (Gliessman, 1990; Bosshard, 2000; Ruiz, 2001).

6. The Avocado Agroecosystem

The avocado (*Persea americana* Mill) is a species native to Mexico and Central America. It belongs to the *Lauraceae* family, which comprises about 2200 species, including woody plants that produce essences that develop in warm climates. It is important to emphasize that the avocado, from the commercial point of view, is the only one that generates foreign currency for the country. There are three horticultural races of avocado - Mexican (*Persea americana* var. drymifolia, M), Guatemalan (*P. americana* var. guatemalensis, G) and Antillean (*P. americana* var. americana, A). These races have environmental conditions differences (drought, salinity, poor soil aeration, etc.) that can affect the development of the area and thus productivity (Schaffer et al., 2013; Wolstenholme,

2013; Salazar-García et al., 2013). A further limitation is an alternation in production (Mickelbart et al., 2012).

Another factor that evaluates the production and longevity of the tree is provoked by *Phytophthora cinnamomi Rands* and a pseudofungus from the Oomycetes Class of the Chromista Kingdom (Castro et al., 2007; Schaffer et al., 2013, Massola and Krugner, 2011). The attack of these pathogens affects the roots of plants of all ages, leading to the death of the plant. It is crucial to prevent the problem by using quality plants, resistant or tolerant genetic materials, and cultural techniques that prevent, minimize and/or control the disease (Smith et al., 2011).

It is worth mentioning that proposals already exist for the conservation and propagation of avocado. However, all plant conservation approaches in our country and worldwide have emerged as a response to the environmental crisis and the massive extinction of species and ecosystems in general (Rivero, 2011). It is necessary to consider that strategies for the preservation of species must be based on at least two perspectives: on the one hand, demographic knowledge of population growth and decline and, on the other, the maintenance of their evolutionary potential (CONABIO, 2000).

Differentiating between *de facto, in situ,* and *ex-situ* preservation of avocado is a resource to indicate that the *de facto* preservation of this agroecosystem refers to longstanding traditions of the indigenous peoples' management or approach of the nature-society relationship (Rivero, 2011). In contrast, Jarvis et al., (2000) define *in situ* preservation as the preservation of ecosystems, natural habitats, and the maintenance and recovery of viable populations of species in their natural environments. On his part, Casas (2011) employs the concept of *ex-situ* to indicate the maintenance of living organisms outside their original and immediate natural environment. In this way, *in situ* and *ex situ* preservation must work together, to establish a comprehensive way of preservation, since it is considered an investment for Mexico's ecological future.

7. Biotechnology and *In Vitro* Cultivation in the Avocado Agroecosystem

Plant tissue culture is a handy tool in biotechnology. This technique is based on "cell totipotentiality," which is the capacity of a plant cell to form a complete plant under certain chemical and physical conditions, given in *in vitro* culture. Thus, the rapid and massive propagation of plants identical to the original is obtained from an isolated part of the plant, whether these are pieces of tissue, meristematic apices, or even isolated cells. Initially, this technique was employed for the massive multiplication of elite plants or micropropagation and also in the obtaining of virus-free clones (Serrano and Piñol, 1991; Huerdo and Merino, 1987). In this way, it is possible to assure a higher yield in the field by the generation of initial material free of pathogens, allowing the exchange of germplasm with other countries.

At present, *in vitro* culture is employed either in the fusion of protoplasts (vegetal cells lacking cellular wall, produced by enzymatic degradation) to obtain hybrids that would not be obtained by traditional methods, or in the culture of anthers to obtain haploids in the development of pure lines (Vidalie, 1986; Adams & Adams, 1991; Brout & Grout, 1995; Copeland & McDonald, 1995). Another alternative use of tissue culture has been the rescue of immature embryos, which allows the combination of variability or the creation of variability in the case that it does not exist, breaking the natural barriers between species and favoring the flow of genes between them or between varieties.

The application of biotechnology through *in vitro* cultivation is an alternative for the *ex situ* preservation of avocado; it allows us to exchange and preserve germplasm and clonal propagation. Finally, the proposed protocols are fundamentally based on micropropagation of different explants to obtain morphogenesis types in avocado (Yasseen, 1993). The *in vitro* preservation of these materials has allowed the preservation of phylogenetic resources for genetic improvement and food security purposes. Thus, organogenesis and somatic embiogenesis have been employed in recent years to obtain a massive propagation free of pests and diseases, with better plant nutrition.

Several authors mention that direct organogenesis has been the most used for cell regeneration. Different researches have been conducted with successful results in the case of different cultivation media, concentrations of growth regulators supplemented with vitamins (Vidales, 2002; Barceló-Muñoz and Pliego-Alfaro, 2003; Nhut et al., 2008; Zulfiqar et al., 2009; Cortés-Rodríguez et al., 2011). In general terms, *in vitro* culture seeks to induce somatic embryogenesis, which is an efficient form of clonal multiplication. This consists of the formation of embryos from somatic cells or tissues, that is, from any cell of the plant, without having to go through the union of the gamma (sexual) cells for the formation of embryos. Furthermore, for genetic engineering, embryogenic cultures are very efficient, since they can produce a large volume of embryonic cells containing genes integrated into the plant genome from a single transformed cell (Rodríguez et al., 1999).

On the other hand, this technique is useful in fundamental studies and the propagation of economically important species; it is also employed in physiological, biochemical, genetic and morphological studies (Rodríguez et al., 1999; McCown & Sellmer, 1987; Pierik, 1990; Angarica & Perea, 1991). Within the tropical fruit trees, they present great world importance in non-industrialized countries such as Brazil, Cuba, Argentina, Turkey, Morocco, and Egypt, due to the wide consumption of their fruits and the possibilities of commercialization. In addition to the advance of communications, it has made it possible for advanced and recently developed technologies to undergo a quite accelerated process of adaptation (Bonga, 1982; Rodríguez et al., 1999).

In contrast, it is possible to observe that the enormous potential that this methodology has caused, in the last 25 years, an increment in the number of tissue culture laboratories in the country for the commercial production of ornamental and fruit plants. *In vitro,* cultivation has been a viable tool for the propagation and establishment of healthy avocado plants. This technique allows maintaining a high production of plants in reduced spaces, under aseptic and controlled temperature and light conditions. However, the current population growth is already overexploiting the Earth's resources. Weising et al., (1995), mentions that humanity must respond to the growing pressures on the Earth's natural resources in order to feed a continuously expanding population.

Since avocado production represents over 30% of the world harvest, using preservation and propagation strategies ensures that this production is not lost.

CONCLUSION

The importance of biotechnology and *in vitro* cultivation in the sustainability of avocado agroecosystem acquires different tonalities depending on the object of study and the context in which it is applied, as, in this sense, the application of biotechnological mechanisms aimed at promoting the sustainable development of avocado through technologies that benefit the environment and are socioeconomically feasible.

REFERENCES

Adams, RP and Adams, J. E. (1991). *Conservation of Plant Genes: DNA Banking and In Vitro Biotechnology.* Academic Press. London, UK.

Angarica, A. and Perea, M. (1991). Tissue culture techniques. CIAT. Cali. Colombia. Cap. 22:494-510. *Avocado Society Yearbook.* The United States of America. 77:101-105.

Barceló-Muñoz, A., and Pliego-Alfaro, F. (2003). Micropropagation of avocado (*Persea americana* Mill). *Forestry Sciences*, 519-542. doi: 10.1007/978-94-010-0125-0_17. Bogota. p. 94.

Bonga, J. M. (1982). *Tissue culture techniques*. P. 4-35 in. Bonga, J. M and Duzan.

Bosshard, A. (2000). A methodology and terminology of sustainability assessment and its perspectives for rural planning. *Agriculture, Ecosystems and Environment*. 77: 29-41.

Brout, B. and Grout, B. (1995). *Genetic Preservation of Plant Cells In Vitro* (Springer Lab. Manual). Springer Verlag. New York. Estados Unidos.

Campbell, C. L., Heck, W. H., Neher, D. A., Munster, M. J. and Hoag, D. L. (1995). Bíophysical measurement of the sustainability of temperate agriculture. In: *The World Bank. Defining and measuring sustainability: The Biogeophysical Foundations*. The World Bank. 252-273.

Casas, R. (2011). *Causes and evidence of soil degradation in the Pampas region*. Buenos Aires: Harvard University.

Castro, M., Fassio, C. and Darrouy, N. (2007). Agronomic evaluation and propagation of new rootstocks and avocado varieties in different agro-climatic zones of Chile. In: *World Avocado Congress*, 6., Viña Del Mar. Proceeding. p. 1-9.

Commission for the Knowledge and Use of Biodiversity.

Conabio. (2000). *Mexico's National Biodiversity Strategy*. Mexico: National.

Copeland, L. O. and McDonald, M. B. (1995). *Principles of Seed Science and Technology. 3rd. Edition.* Chapman & Hall. NY, Estados Unidos.

Corcos, A. F., Monaghan, F. V. and Weber, M. C. (1993). *Gregor Mendel's Experiments on Plant Hybrids: A Guided Study* (Masterworks of Discovery). Rutgers Univ Press, NJ, Estados Unidos.

Cortés-Rodríguez, M. A., López-Gómez, R., Martínez-Pacheco, M. M., Suárez-Rodríguez, L. M., Hernández-García, A. and Salgado-Garciglia, R. (2011). In vitro propagation of Mexican race avocado (*Persea americana* Mill var. drymifolia). *Acta Hort*. 923:47-52.

DJ (ed) *Tissue Culture in Forestry*. Martinus Nijhoff/Dr. W. Junki Publichers Printed in the Netherlands.

Dasilva, E. J. (2004). *The Colours of Biotechnology: Science, Development and*.

Dreher, K., Morris, M., Khairallah, M., Ribaut, M., Pandey, J. M. and Srinivassan, S. G. (2000). Is marker-assisted selection cost-effective compared to conventional plant breeding methods—the case of quality protein maize. Paper presented at the *4th ICABR Conference on Economics of Agricultural Biotechnology*, Ravello, Italia, 24-28 August.

Freeze, FY. (2002). *Feeding or fooling the world Can GM really feed the world.*

Gliessman, S. R. (1990). Agroecology. Researching the ecological basis for sustainable agriculture. Introduction. In: Gliessman, S. R. (ed). Agroecology. Researching the ecological basis for sustainable agriculture. *Ecological Studies,* 78. Springer-Verlag. New York, Berlin, Heidelberg, London, Paris, Tokyo and Hong: 3-29.

Gomez, A. L. (1999). *Biotechnology as a tool for the conservation and use of cultivated diversity Plant Biotechnology Laboratory*. Centre for Agronomic Research. The University of Costa Rica.

Gomez, C. H., Benavides, C. P., Osorio, D. M. and Cuervo, M. A. (2013). *Strategic Action Plan for the Bio Sector of Manizales*. Foundation University Company State of the Axis Coffee.

Gonzalez, G. L. (1999). *Towards the strengthening of national capacity for the adequate management of cultivated biodiversity Conarefi*. IICA. San Jose.

Guarino, Rao L. V. and Reid, R. (1995). *Collecting Plant Genetic Diversity: Technical Guidelines.* CAB International, Reino Unido.

Howell, SH (1998). *Molecular Genetics of Plant Development*. Cambridge Univ. Press. MA., Estados Unidos.

Humankind. *Electronic Journal of Biotechnology*, 7(3), 01-02.

Hurtado, D. V. and Merino, M. E. (1987). *Plant tissue culture*. Editorial Trillas, Mexico.

Jarvis, D., Myer, L., Klemick, H., Guarino, L., Smale, M., Brown, A., Sadiki, M., Sthapit, B., and Hodgkin, T. (2000). *A training guide for in situ conservation on farm. Version 1.* Rome, Italy: International plant genetic resources Institute.

Kelly, A. F., and George, R. A. (1998). *Encyclopedia of Seed Production of World Crops*, John Wiley & Sons. NY. Estados Unidos.

Kjellsson, G., Somonsen, V., and Ammann, K. (1997). *Methods for Risk Assessment of Transgenic Plants: Ii, Pollination, Gene-Transfer, and Population Impacts.* Birkhauser. MA. The USA.

Lee, H. (1994). Sustainable agriculture systems, eds J. L. Hatfield and D. L. Karlen. 316 pp. Boca Raton: CRC Press (1994) ISBN 1566700493. *The Journal of agricultural science*, 123.(3), 422-422. Doi: 1 0.1017/ S0021859600070568.

Marx, JL (1989). *Revolution in Biotechnology*. ICSU Press England.

Massola, NS, and Krugner, T. L. (2011). Fungos Phytopathogenic. In: Amorin, L., Rezende, JAM, and Bergamin, F. A. (Ed.). *Phytopathology Handbook. 4. ed.* Piracicaba: Agronomic Ceres. p. 149-206.

MC Cownn, BH, and Sellmer, C. (1987). In J. M. Bonga; Durzan. Cell and *Tissue Cultura in Forestry Vol. 1.* Mains general and biotechnology Martinus Nijhoff publisher.

Mickelbart, M. V., Robinson, P. W., Witney, G., and Arpaia, M. L. (2012). 'Hass' avocado tree growth on four rootstocks in California. II. Shoot and root growth. *Scientia Horticulturae* 143, 205–210.

Mohan Jain, S. M., Brar, D. S. and Ahloowalia, B. S (1998). Somaclonal Variation and Induced Mutations in Crop Improvement (*Current Plant Science and Biotechnology in AgricuLture, No 32*). Kluwer Academic Pub. Londres, Reino Unido.

Narvaez, A. (2015). The biotechnology market in Colombia. Spanish Embassy in Nhut, D. T., Thi, N. N., Khiet, B. L. T. and Luan, VQ. (2008). Peptone stimulates in vitro shoot and root regeneration of avocado (*Persea americana* Mill). *Scientia Horticulturae*. 115:124-128.

Paszkowski, J. (1994). *Homologous Recombination and Gene Silencing in Plants*. Kluwer Academic Pub. Londres, Reino Unido.

Pérez, J. N. (1998). Propagation and plant breeding by plant biotechnology. *5th International Colloquium on Plant Biotechnology*

Institute of Plant Biotechnology. Book of reports. Santa Clara, Cuba. pp. 25-31.

Pierik, R. M. L. (1990). *In vitro cultivation of higher plants. Third edition.* Editions World Press, 320 p.

Potrykus, I. 2003. From "golden" to "nutritionally optimized" rice – and from a scientific concept to the farmer. Presentation delivered at the conference "*In the Wake of the Double Helix: from the Green Revolution to the Gene Revolution*," Bologna, Italy, 27–31 May.

Querol, D. (1993). *Genetic Resources: A Practical Guide to Their Conservation.* Zed Books. Londres, Reino Unido.

Rivero, M. (2011). Plant tissue culture for conservation. *Polybotany* 3:10-21.

Rodriguez, N. N., Capote, M., and Zamora, M. (1999). In vitro culture of the avocado tree (*Persea americana* Mill). Research Institute of Citrus and other Fruit Trees, Cuba. *Chapingo Journal Horticulture* S. 5:231-237.

Ruiz, R. O. (2001). *Agroecological sustainability in Kent. England: the systems approach at catchment and parish group levels*. Ph.D. Thesis. Imperial College, University of London. pp. 322.

Salazar-Garcia, S., Garner, L. C., and Lovatt, CJ (2013). Reproductive Biology. In: Schaffer, B. A., Wolstenholme, B. N. and Whiley, AW (Ed.). *The Avocado: Botany, Productions and Uses. 2nd ed.* Homestead: CABI, 6 (2) p. 118-167.

Schaffer, B. A., Gil, P. M., Mickelbart, M. V. and Whiley, AW Ecophysiology. In: Schaffer, B. A., Wolstenholme, B. N. and Whiley, AW (Ed.). (2013). *The Avocado: Botany, Productions and Uses. 2nd ed.* Homestead: CABI. Cap. 7, p. 168-199.

Serrano, M., and. Piñol, M. T. (1991). *Plant biotechnology. Ed.* Synthesis, Madrid. 285p.

Shmaefsky, B. R. (2013). Agricultural Biotechnology: History, Science, and Society. Bibliographic Essay. and Society. *Bibliographic Essay.* CHOISE. p. 199-208.

Smith, L. A., Dann, E. K., Pegg, K. G., and Whiley, AW (2011). Field Assessment of Avocado Rootstock Selections for Resistance to

Phytophthora Root Rot. *Australasian Plant Pathology*, Collingwood, v. 40, n. 1, p. 39-47.

Sobral, BWS (1996). *The Impact of Plant Molecular Genetics*. Birkhauser. MA, United States.

Vidales, FI (2002). *Effect of growth regulators on avocado organogenesis and somatic embryogenesis processes* (*Persea americana* Mill.). Thesis. University of Colima, Tecoman, Colima, Mexico. 62-119.

Vidalie, H. (1986). *In vitro culture*. Scientific Publisher, Mexico.

Weisign, K., Nybom, H., Wolff, K., and Meyer, E. (1995). *DNA Fingerprinting in Plants and Fungi*. Book Britain, London, UK.

Wolstenholme, BN (2013). Ecology: climate and soils. In: B. A. Schaffer; Wolstenholme, B. N., Whiley, AW (Ed.). *The Avocado: Botany, Productions and Uses. 2nd ed.* Homestead: CABI. Chap. (52)P. 86-117.

Yasseen, M. Y. (1993). *Morphogenesis of avocado in vitro*. A review. California.

Zulfiqar, B., Abbasi, N. A., Ahmad, T., and Hafiz, I. A. (2009). Effect of Explant Sources and Different Concentrations of Plant Growth Regulators on In Vitro Shoot Proliferation and Rooting of Avocado (*Persea americana* Mill.) cv. "Fuerte." *Pak. J. Bot*. 41(5):2333-2346.

In: Avocado: Consumption and Health ISBN: 978-1-53618-204-0
Editors: M. G. Melgar Lalanne et al. © 2020 Nova Science Publishers, Inc.

Chapter 6

Patents on Avocado Products Related to Health

Alicia Ortiz-Moreno[1,*], Lidia Dorantes-Alvarez[1], Hugo Necoechea-Mondragon[2] and Nancy Ordaz-Trinidad[1]

[1]Instituto Politécnico Nacional, Escuela Nacional de Ciencias Biológicas, Departamento de Ingeniería Bioquímica, Unidad Profesional López Mateos, Ciudad de México, México
[2]Instituto Politécnico Nacional, Coordinación de Operación de Redes de Investigación. Miguel Bernard s/n Colonia La Escalera, México City, México

Abstract

This chapter contributes to the revision of patents of avocado products related to health. *Persea americana*, commonly known as avocado, belongs to the Lauraceae family that is native to Mexico and Central America and is currently cultivated in most tropical and

* Corresponding Author's Email: ortizalicia@hotmail.com.

subtropical countries of the world. Avocado is an economically important crop due to its nutritional value; however, the three constituents, the fruit peel, seed, and leaves, are underexploited.

Patents were obtained when the infusions or hydroalcoholic extracts from avocado leaves, seeds, and peel were shown to have pharmacological activities and that some of them have analgesic, antimicrobial, antidiabetic, hypoglycemic, and anti-inflammatory properties. Avocado seeds can be used to obtain avocado oil enriched in alkyl polyols and acetylated derivatives thereof for use as a drug advantageously in the prevention and treatment of conjunctive tissue disorders, such as arthrosis, articular pathologies such as rheumatism, or periodontal diseases such as gingivitis or periodontitis.

There are several patents related to the high content of avocado polyphenols, which are extracted for use in preventing or treating disorders or pathological conditions of the skin, the mucous membranes, or superficial body growths.

The fruit of avocado contains mannoheptulose, perseitol, various vitamins, abundant fat and protein, and high contents of sodium, potassium, magnesium, and calcium. This content is used for treating or preventing diseases related to the modification of innate, acquired immunity, and treating or preventing overweight or obese in mammals.

Keywords: avocado, health, patents, byproducts

1. Introduction

Avocado (*Persea americana* Mill.) is a tropical and subtropical fruit that is native to Mexico and Central America. Avocado is gaining increasing worldwide acceptance and has received extensive marketing and a wide distribution due to its essential nutritional benefits for human health. Mexico harvests more than 30% of the avocados worldwide, representing the leading producer and exporter of avocados, which has become a crop of high interest, and this has a great economic impact on Mexico [1]. Different varieties of avocados are cultivated worldwide, and Hass and Fuerte are among the most consumed [2].

Table 1. Recent patents on avocado products related to health

Patent	Subject	Application	Year	Ref.
WO 2012042404 A2	*Persea* spp. (avocado) enriched in bioactive compounds that have been used as antimicrobial, antibacterial, or spore germination is inhibiting agents, the process for obtaining the extracts, acetogenins, and isolated molecules and methods for using the extracts enriched in bioactive compounds for providing antimicrobial, antibacterial or spore germination.	Antimicrobial agent	2012	11
US20130216488A1	*Persea* spp. (avocado) enriched in bioactive compounds that have been used as antimicrobial, antibacterial, or spore germination inhibiting agents, the process for obtaining the extracts, acetogenins, and isolated molecules and methods for using the extracts enriched in bioactive compounds for providing antimicrobial, antibacterial or spore germination.	Antimicrobial agent	2013	12
EP 2 851 062 A2	*Persea* spp. (avocado) enriched in bioactive compounds that have been used as antimicrobial, antibacterial, or spore germination inhibiting agents, the process for obtaining the extracts, acetogenins, and isolated molecules and methods for using the extracts enriched in bioactive compounds for providing antimicrobial, antibacterial or spore germination.	Antimicrobial agent	2015	13
US 004.0022882A1	Producing an avocado leaf extract rich in furanic lipids, the raw material for which consists of avocado leaves, and which comprises the following steps: a liquid-solid extraction step followed by evaporation under vacuum, and heat treatment these two steps can be carried out in this order or the reverse order.	Antimicrobial agent	2004	14
USOO9421234B2	Orally administrable compositions comprising avocado/Soybean unsaponifiables and lipoic acid or derivatives thereof.	Anti-inflammatory agent	2016	15
WO2005115422A2	Tea containing *Persea americana* var. drymifolia leaf (0.2 mg/mL to about 0.4 mg/mL). A dietary supplement composition for lowering cholesterol levels in mammals comprising a *Persea americana* var. drymifolia leaf extract and a carrier wherein the extract, consists essentially of a water-soluble fraction.	Cardiovascular disease risk reduction	2005	16

Table 1. (Continued)

Patent	Subject	Application	Year	Ref.
USOO6440464B1	The present invention relates to the use of nutritional supplement compositions to overcome nutritional deficiencies typically associated with the normal mammalian diet. The compositions of the present invention are obtained by combining fish oil, garlic powder, rutin, capsaicin, Vitamin A, Vitamin C, Vitamin E, Selenium, and one or more juice concentrates.	Cardiovascular disease risk reduction	2002	17
USOO902381 OB2	The use of a compound comprising D-mannoheptulose and/or perseitol for producing a drug for treating and/or preventing diseases related to the modification of an innate and/or acquired immunity by increasing the production of antimicrobial peptides, preferably hBD-2 without inducing inflammatory reactions, irritation or intolerance. Said compound can also comprise a peptidic avocado extract and/or a peptidic lupin extract.	Strengthens innate immunity and anticancer agent	2015	18
WO2015070252A1	Process for preparing a carbohydrate extract comprising mannoheptulose from avocados, as well as compositions, including food compositions, comprising such extracts. Also described herein are methods for using the carbohydrate extract comprising mannoheptulose and compositions comprising the carbohydrate extract comprising mannoheptulose.	Strengthens innate immunity and anticancer agent	2014	19
US20140348986A1	Avocados are processed in a controlled manner to optimize and maintain their levels of mannoheptulose and perseitol. The avocado flesh is used as a source of these materials in food compositions, especially pet foods.	Strengthens innate immunity and anticancer agent	2014	20
WO2014190136A1	The invention relates to the use of criollo avocado in the manufacture of orally ingestible compositions is used to prepare nutritional compositions for oral ingestion by humans and lower animals, especially pets such as dogs and cats.	Strengthens innate immunity and anticancer agent	2014	21
WO2007057439A1	A process for obtaining a triglyceride-rich refined avocado oil. Advantageously, the refined avocado oil, according to the invention, contains a sterol-enriched unsaponifiable fraction. Advantageously, the refined avocado oil, according to the invention, is substantially free of acetogenins and furanic lipids. Such compositions for their use as a medicament, as a dermatological agent, as a cosmetic agent or as a nutraceutical agent, to be used humans or animals.	Dermatological agent	2007	22

Patent	Subject	Application	Year	Ref.
US20130183255A1	The invention relates to an avocado extract rich in polyphenols, an extract for use thereof in preventing or treating disorders or pathological conditions of the skin, the mucous membranes, or superficial body growths. Finally, the invention relates to a cosmetic care process for the skin, superficial body growths, or mucous membranes, to improve the condition or the appearance, which consists of administering such a composition or such an extract.	Dermatological agent	2013	23
EP2654766B1	A polyphenol-rich extract of avocado fruit, which is at least 30% procyanidins. A composition including an active agent characterized as having a cosmetic, pharmaceutical, dermatological, or nutraceutical composition. Further, it includes another active agent, antioxidant agents, antiaging agents, depigmenting or hypopigmented agents, pigmenting agents, lipolytic or lipogenesis inhibitor agents or anti-cellulitis or slimming agents, organic or mineral sunscreens and filters (pigmentary or ultrafine), antifungal compounds, preservatives, antibacterial agents, prebiotics and probiotics, antibiotics, and immunomodulators.	Dermatological agent	2012	24
US6440437B1	The invention particularly relates to wet wipes and wipe-type products that can be used for a variety of purposes, including cleaning, cosmetics removal, and sanitation, comprising an oil-in-water emulsion composition. A solution that cleans the surface of the skin and provides enhanced skin health benefits.	Dermatological agent	2002	25
US6287581B1	A superior skin barrier was enhancing body facing material, a lipid-enriched hydrophobic composition comprising a natural fat or oil, a sterol or sterol derivative, an emulsifying surfactant, a humectant, an emollient, a wax, and a viscosity enhancer, and after that resolidifying the composition to form a distribution of solid composition on the outer surface of the body facing material.	Dermatological agent	2001	26

Avocado fruits have high nutritional quality and contain high levels of vitamins, minerals, proteins, and fibers, as well as high concentrations of unsaturated fatty acids, which are beneficial to health [3]. Also, avocado peel and seeds have high contents of bioactive phytochemicals such as phenolic acids, condensed tannins, and flavonoids, including procyanidins, flavonols, hydroxybenzoic, and hydroxycinnamic acids [4, 5, 6]. These bioactive compounds have shown various biological activities, such as antioxidant and anti-inflammatory properties [7].

Avocado industrial processing generates large quantities of agro-industrial byproducts, mainly peel and seed, ranging from 18% to 23% of the fruit dry weight [8]. Thus, it is interesting to reuse these byproducts both to reduce their negative impact on the environment and to add value to them, in as much as they are sources of critical phytochemical compounds [9, 10]. Therefore, the transformation of these agro-industrial byproducts, which yield a range of bioactive substances, should be made by green extraction to reduce the environmental impact of the agro-industrial chain. In this chapter, we describe patent applications for avocados, with an emphasis on its byproducts, focusing on the proper use of waste and the possibility of monetizing waste for health purposes. The whole avocado is rich in bio compounds (pulp, seed, and peel) and has many health benefits, such as antimicrobial, antioxidant, and cardiovascular disease risk reduction, as well as dermatological uses, among others. In this chapter, we demonstrate the current panorama of byproduct patents of avocado byproducts (Table 1).

2. Antimicrobial Agent

Avocado leaf extracts show antibacterial and antifungal activity [11, 12, 13], which makes it particularly attractive as a new natural antimicrobial compound in response to the growing demand for safe and healthy foods, using natural preservatives. Different authors have developed avocado leaf-based products with antimicrobial activity.

Extracts from *Persea spp.* (avocado) that are enriched in bioactive compounds have been used as antimicrobial, antibacterial, or spore germination inhibiting agents. The process for obtaining the extracts, acetogenins, and isolated molecules and methods for using the extracts enriched in bioactive compounds for providing antimicrobial, antibacterial, or spore germination inhibiting effects have been reported by Hernandez-Brenes [11, 12, 13].

Extracts enriched in naturally occurring acetogenins from Persea spp. are characterized by having antimicrobial, antibacterial, or spore germination inhibiting effects. An acetogenin-enriched extract that presents spore germination inhibiting activity, as a sporicidal and sporostatic agent against native bacterial spores from *Clostridium spp., Bacillus spp.* Furthermore, *Alicyclobacillus spp.,* among other pathogenic and non-pathogenic bacteria, has been found [11, 12, 13]. The extract could be used in pharmaceuticals, foods, personal care, and cleaning compositions or products comprising the said extract, thus having antimicrobial, antibacterial, or spore germination inhibiting effects. The enriched extract is useful as an antimicrobial agent to inhibit the growth of viable cells of other non-spore-forming gram-positive bacteria, such as *Listeria monocytogenes*, in combination with refrigerated conditions [11, 12, 13].

Given the therapeutic value of avocado furanic lipids for their beneficial and curative action on the connective tissue, in particular in inflammatory pathologies such as osteoarthritis, periodontitis, and scleroderma, and their high cost in general, a keen interest exists in finding alternative routes for preparing these furanic avocado lipids. Piccirilli and Legrand (2004) developed a method that makes it possible to obtain an avocado leaf plant extract rich in furanic lipids, namely, having a content ranging from 20% to 80%, and preferably 30% to 50% [14]. The avocado leaves may be obtained from avocado plants belonging to the following varieties:

Hass, Fuerte, Ettinger, Bacon, Nabal, Anaheim, Lula, Reed, Zutano, Queen, Criola Selva, Mexicana Canta, Region Dschang, Hall, Booth, Peterson, Collinson Red or Alpha Krome and more particularly to the varieties Hass, Fuerte and Reed [14].

3. Anti-Inflammatory Agent

The damage to connective tissue may be a result of physical injury or may represent "wear and tear" from regular use, weight and age, for example, from osteoarthritis. Damage to connective tissue may also result from diseases such as rheumatoid arthritis, synovial disorders, infection-related rheumatic diseases, and inflammatory connective tissue disorders.

Many studies have documented the benefits of avocado/soybean unsaponifiables (ASU) for promoting joint health and the management of osteoarthritis. Clinical studies have reported beneficial effects of ASU in human and equine osteoarthritis patients as well as in experimental animal models of OA. The mechanisms that could account for the beneficial effects of ASU for osteoarthritis have been studied in vitro using bovine and human joint tissue cells. These studies have shown that ASU inhibits the expression and production of cytokines, chemokines, PGE, nitric oxide, and MMPs. ASU also exerts anabolic effects on cartilage metabolism by enhancing the synthesis of cartilage matrix components while suppressing their degradation [15].

Frondoza et al. (2016) provide a method for preventing, treating, protecting, repairing or reducing damage to connective tissues or reducing symptoms associated with damage to connective tissue in an avian or mammalian subject comprising administration of the following to the subject: (i) avocado/soybean unsaponifiables, and (ii) lipoic acid, or derivatives thereof.

The term "avocado/soybean unsaponifiables (ASU)" refers to a mixture of avocado unsaponifiables and soybean unsaponifiables. "Unsaponifiables" are compounds that do not react with alkali to form a soap. The term "avocado unsaponifiables" refers to an extract of compounds obtained from any part of an avocado (genus Persea). The avocado may be any species of avocado, such as, but not limited to, *Persea americana* and *Persea schiedeana* [15].

4. CARDIOVASCULAR DISEASE RISK REDUCTION

High cholesterol generally means that a person's total blood cholesterol level is more than 240 mg/dL or that a person's low-density lipoprotein level is more than 160 mg/dL. One common intervention is a class of lipid-regulating pharmaceuticals called statins. A popular statin is LIPITOR®. Although these agents have been proven safe in clinical trials, like any drug, they carry the risk of undesirable side-effects.

It has now been found that drymifolia leaves contain agents that lower total cholesterol levels [16]. Accordingly, compositions containing drymifolia leaf and drymifolia leaf extracts are believed to be natural alternatives to LIPITOR® and other prescription drugs.

Broverlli et al. (2005) developed a cholesterol-lowering composition containing drymifolia leaf and drymifolia leaf extracts. In one procedure, the drymifolia leaves are dehydrated, milled, and prepared for consumption as a tea. Surprisingly, the leaves of drymifolia showed a cholesterol-lowering effect comparable to LIPITOR® and a much higher cholesterol-lowering effect than the fruit and the leaves of other avocado varieties commonly grown in North America, namely, Persea nubigena var. guatamalensis cv. Nabal and Persea nubigena var. guatamalensis cv. Haas. In another process, compositions, such as dietary supplements, are provided containing an extract prepared from dehydrated drymifolia leaves. In a third process, a composition comprising drymifolia leaf extract in combination with other cholesterol-lowering agents is provided. In a fourth process, a method of lowering cholesterol levels in a mammal by administering a drymifolia leaf extract is provided [16]. The particular substance(s) in drymifolia leaves that cause cholesterol-lowering is not yet understood. However, the leaves prepared as tea and extracts derived from the leaves show cholesterol-lowering activity.

In the last few years, scientific literature has provided strong evidence of the correlation between micronutrient compositions (such as vitamins, minerals, fish oils, and plant extracts) and cardiovascular disease. For humans with a high risk of cardiovascular illnesses, realizing a significant reduction in the levels of high cholesterol, triglycerides, and low-density lipoprotein (LDL) in their blood serum is known to be important for reducing the risk of cardiac diseases. It is also known that affecting an increase in the levels of high-density lipoprotein (HDL) also provides a significant decrease in the risk of cardiac diseases [17].

Cardiovascular disease resulting from the buildup of arterial plaque is known to be a leading cause of illness and death in humans. Arterial plaque is caused by precipitous material formed chiefly of oxidized low-density lipoprotein (O-LDL). The buildup of plaque in the form of O-LDL in the arteries is understood to be a factor in ischemic heart disease. Free radical oxidants, many of which come from naturally occurring sources such as sun exposure, metabolism of certain nutrients, exercise, or are otherwise often observed in persons who have diabetes and high blood pressure, act to oxidize LDL into its deleterious forms O-LDL. Free radical "scavengers" such as vitamins A, E, C, and selenium are believed to react with these oxidants and render them incapable of oxidation. The inhibitory action of these antioxidants thus inhibits the formation of O-LDL, thereby lowering the levels of arterial plaque deposits in blood vessels.

Although compositions used to reduce the risks of cardiovascular disease are known, Hsia and Fan (2002) developed a novel combination of fish oil, garlic, rutin, capsaicin, Vitamin A, Vitamin C, Vitamin E and selenium, and juice concentrates (including avocado, 0.5% to about 20% of the weight of the total juice concentrate), which may be used to treat or prevent cardiovascular disease and disorders [17].

5. Innate Immunity Strengthening and Anticancer Agents

All animal species are confronted daily with a large number of microorganisms, such as bacteria, fungi, parasites, or viruses, that can threaten their health or even their life. Two systems of defense oppose these microorganisms: a system called innate immunity, which is common to all animals, including man, and a so-called adaptive or specific immune system that is acquired thanks to cells and immunity mediators after contact with the potential aggressor.

The innate and adaptive immunity responses differ in their recognition mechanisms for these microorganisms. In innate immunity, the specificity of the receptors is genetically determined from birth and does not change. These receptors are expressed on the cells, such as specific epithelial and endothelial cells, dendritic cells, monocytes, and macrophages. All the structures recognized by the innate immunity receptors are common to a very significant number of microorganisms. In contrast to the adaptive immunity response, the innate immunity response mechanisms (phagocytosis, antimicrobial peptides, etc.) are activated at the onset of the infection and control almost immediately the proliferation of pathogens that invade the host.

Antimicrobial peptides have been found in both the vegetal and animal kingdoms, and more than 500 antimicrobial peptides have been discovered from insects to man. The antimicrobial peptides are small molecules (10 to 50 amino acids) capable of destroying a wide variety of microorganisms (Gram+ and/or Gram-, fungi, viruses, and transformed cells) by rendering their cellular membrane permeable. Moreover, some of these antimicrobial peptides, through their chemoattractive properties, can recruit the cells participating in adaptive immunity, such as dendritic cells, or still more, the T lymphocytes. Numerous antimicrobial peptides have been found in the vernix caseosa, in the amniotic fluid, and the skin of the newborn, suggesting their crucial role in the antimicrobial defense during birth but also from the beginning of life while the acquired immunity is still immature.

Piccirilli A. et al. (2015) developed a drug for protecting the healthy or pathological skin of adults or the elderly, notably of immunodepressive individuals. The drug is also suitable for encouraging cicatrization in the process of normal or pathological cicatrization, such as ulcers and scabs; and the drug is aimed at the treatment and/or prevention of periodontal diseases, inflammatory articular pathologies such as arthritis, mucous infections, notably of the vaginal, intestinal, respiratory, nasal or auricular mucosa, or infections of the visual system.

The source of D-mannoheptulose and/or perseitol can be a hydrosoluble extract of sugars from avocado or another plant. Otherwise, D-mannoheptulose and perseitol are commercially available (synthetic origin). According to an advantageous variant, the source of D-mannoheptulose and/or perseitol is a hydrosoluble extract of sugars from avocado [18].

The hydrosoluble extract of sugars from avocado can be directly obtained from any part of the avocado or the avocado tree, such as the fruit, the skin, the stone, the leaf, or the roots of the avocado tree. It is also possible to obtain a peptidic extract from the coproducts of the avocado conversion industry, among which the following can be cited, among others: fresh avocado pulp, frozen pulp, dehydrated pulp, avocado oil cake obtained from oil extraction processes (mechanical and/or solvent extraction of previously dehydrated fruit), de-oiled solids resulting from wet oil-extraction processes (centrifugation processes), de-oiled solids resulting from enzymatic extraction processes of avocado, raw avocado mash (guacamole), and solid waste resulting from production units of these mashes. The extract is advantageously obtained from the fresh fruit of the avocado tree. The fruits may be chosen from among the varieties Hass, Fuerte, Ettinger, Bacon, Nabal, Anaheim, Lula, Reed, Zutano, Queen, Criola Selva, Mexicana Canta, Region Dschang, Hall, Booth, Peterson, Collinson Red, and more advantageously among the varieties Hass, Fuerte, and Reed. The varieties Hass, Fuerte, Ettinger, and Bacon are preferred and are more advantageous than the Hass and Fuerte varieties [18].

The highest content of mannoheptulose and/or perseitol, another seven-carbon sugar, has been reported in avocados. Mannoheptulose and perseitol are unstable in avocados and decline rapidly as the fruit ripens. Interconversion of mannoheptulose and perseitol has been reported; perseitol can be oxidized to mannoheptulose by enzymes present in the extract of the fruit. Martinez-Serna, M. (2015) describes carbohydrate extracts from avocados that contain unusually high amounts of mannoheptulose and its related seven-carbon sugar alcohol perseitol, together comprising approximately 50-90% of the total soluble sugars [19]. Other six-carbon sugars present in the extracts include glucose, fructose, and sucrose. Mannoheptulose is of interest because it is a classical inhibitor of glucose-induced insulin secretion and glucose oxidation.

Mannoheptulose inhibits glucose-induced insulin secretion by selectively inhibiting the enzyme glucokinase. By blocking glucose phosphorylation, the breakdown of glucose is inhibited. Mannoheptulose has also been implicated as an anticancer agent, most likely due to its ability to inhibit cell growth in cell types expressing glucokinase, such as in liver tumor cells. Additionally, mannoheptulose has been described as an antioxidant. As such, mannoheptulose has been implicated as having a variety of effects on mammalian metabolism and health.

Martinez-Serna improved the process of obtaining mannoheptulose, with higher yields of mannoheptulose without compromising the integrity of mannoheptulose [19]. This enhanced process may also generate, as byproducts, avocado oil, and avocado solids that are rich in vitamins A, B, D, E, lutein, carotenoids, and/or proteins that are useful in health and beauty care formulations.

Beyer and Ter Haar (2014) determined that, from an economic standpoint, it would be quite advantageous to be able to use the avocado fruit, especially the peeled, depitted flesh of the fruit, as a natural source of mannoheptulose and perseitol [20, 21]. However, many varieties of avocado have been bred, not for their mannoheptulose/perseitol levels, but rather to provide a high oil content. Moreover, avocados can lose a considerable amount of their mannoheptulose during fruit ripening or even during postharvest storage. Unfortunately, using unripened avocado fruit as a source of mannoheptulose and/or perseitol is quite problematic, due to the comparative toughness of the fruit's unripened flesh. Removing the pit from the unripened fruit also presents a significant problem for the formulator since it is held tenaciously by the tough avocado flesh. The peel is also challenging to remove. Of course, it would be optimal to employ avocados that are not only in abundant supply but also have the highest possible concentration of mannoheptulose and perseitol. The most common and abundant species of avocado that is grown in most orchards in the Northern Hemisphere is the Hass. Unfortunately, it has been determined that, while noted for its flavor and oil content, the Hass avocado is relatively low in mannoheptulose content compared with the West Indian and West Indian/Guatemalan hybrid varieties. Accordingly, those latter varieties are preferred for use.

The process developed by Beyer and Ter Haar (2014) provides avocados that have been optimized concerning both mannoheptulose/ perseitol levels and to the removal of the pit and the peel from the fruit to give the desired avocado flesh [20, 21].

6. Dermatological Agent

The avocado is originally from Mexico and was brought back to Europe by the Spanish conquistadors. This fruit has become a staple food. The avocado is rich in oil since the lipids represent between 10% and 20% of the fruit fresh. The great richness in fat is indeed one of the characteristics of this fruit. The oil, which by its fatty acid profile, is very close to that of olive, has been used for many years for its applications in cosmetics and pharmacy. Besides, thanks to its nutritional properties, its interest as a food oil or supplement has appeared in recent years, and global production is steadily progressing.

Raw or virgin avocado oils are usually extracted by a process qualifying as cold pressure. When using conventional refining processes, including degumming, neutralization, discoloration, fragmentation, and deodorization, the different treatments are not efficient enough to guarantee the elimination of total unstable compounds or their non degradation as secondary derivatives. Thus, the low stability of some constituents of the lawyer and the risks associated with their potential toxicity requires control over the fate of these constituents to produce a food-grade oil. Msika, Philippe, and Legrand, Jacques (2007) developed a new method to obtain refined avocado oils that did not have the disadvantages of the prior method and to market dried avocado oils [22] safely.

The refined avocado oil or the compositions, according to Msika, Philippe, and Legrand, Jacques, can also be used in cosmetic compositions intended for the treatment, repair, or protection of the skin or dry mucous membranes, as well as for dry or dull hair. The oil or compositions may also be used for the cosmetic treatment of disorders related to aging of the skin, neighboring mucous membranes and/or superficial body growths; for treatment of skin disorders, adjacent mucous membranes and/or integuments resulting from exposure to actinic radiation, in particular, UV radiation; for treatment of diseases of the skin, neighboring mucous membranes and/or integuments resulting from exogenous stress (pollution, etc.) or endogenous stress; or as agent softeners, regenerating or strengthening the hair. The refined avocado oil or the compositions may also be used as a drug for repairing tissue or the treatment of inflammation or irritation of the skin, integuments or mucous membranes, especially to treat eczema, psoriasis, pruritus ichthyosis, acne, xerosis, atopic dermatitis, atopic skin or even skin allergies [22].

Saunois, Alex et al. (2013) reported that polyphenol-rich extracts of avocado (*Persea gratissima* or *americana*) have cosmetic, dermatological, pharmaceutical, and nutraceutical properties, in particular, cosmetic and dermatological properties never described to date [23, 24]. In particular, it is the first time that polyphenol-rich extracts of avocado have been used as such for their specific properties.

The cosmetic method is also characterized by the fact that the composition or extract is used as a moisturizing product or as a chronological or photo-induced antiaging product in the prevention of aging or photo-induced aging, as a slimming and/or anti-cellulitis product, or as an antiaging or anti-age spot product [23, 24].

Advantageously, the composition or extract, according to Saunois, Alex et al. (2013), may be used for the prevention and/or treatment of reactions, disorders or pathologies of the skin, such as rosacea or erythrocouperosis, psoriasis, vascular diseases, diaper rash, atopic dermatitis, eczema, contact dermatitis, irritative dermatitis, allergic dermatitis, seborrheic dermatitis (cradle cap), sensitive skin, reactive skin, pruritus, dry skin (Xerosis), dehydrated skin, skin with redness, cutaneous erythema, aged or photoaged skin, photosensitive skin, pigmented skin (melasma, postinflammatory pigmentation), skin with cellulitis, loose skin,

skin with stretch marks, scurf, chapping, bites, cracks, in particular of the breasts, sunburn, inflammation due to rays of all types, irritation by chemical, physical (for example, tension stress for pregnant women), bacteriological, fungal or viral, parasitic (lice, scabies, tinea, mites, dermatophytes) or radiological agents, or by a deficit in innate immunity (antimicrobial peptides) or acquired immunity (cellular, humoral, cytokines), and/or mucous membranes such as the gums and periodontium subject to gingivitis (sensitive gums of newborns, problems of hygiene, due to the use of tobacco or other products), periodontopathies, or genital mucosa presenting irritation of the external or internal male or female genitalia, and/or epithelial appendages such as the nails (breakable, fragile nails, etc.) and hair (alopecia, dandruff, hirsutism, seborrheic dermatitis, folliculitis) whether immature, healthy or mature, presenting in particular disorders of the scalp such as androgenetic, acute, localized, cicatricial and congenital, occipital in infants and greata alopecia (or pelade), due to chemotherapy/radiotherapy or telogen effluvium, anagen effluvium, pilar dystrophy, trichotillomania, tinea or oily or dry dandruff [23, 24].

The stratum corneum is the outermost layer of the skin and is responsible for regulating skin water levels. It also functions as a barrier against chemicals and other stressors found in the environment. The complex arrangement of lipids in the intercellular space of the stratum corneum is responsible for the establishment of normal barrier function.

Multilayered structures of cholesterol, ceramides, and fatty acids, as well as some other minor lipids, provide a significant barrier to the transport of hydrophilic substances into or through the skin. The link between the barrier function and skin health can be observed through skin inflammation that results from the extraction of lipids from the skin [25, 26].

Diapered skin is typically cleaned by the application of cotton washcloths or disposable wet wipes. The effect on the skin of the wet wipes depends on the materials and surfactant systems used for the wipes. Diapered skin is frequently cleansed several times a day with wipes utilizing solutions containing surfactants. Surfactants are known to extract lipids from the stratum corneum or disorganize the lipid structure within the stratum corneum, thereby decreasing the barrier function. The wet wipe and wipe-type product material can cause physical abrasion damage to the skin and can lead to an increase in transepidermal water loss, an indication of decreased barrier function. Other instances from which skin barrier function interruption results are frequent hand washing and contact dermatitis, originating from harsh surfactants and other chemical irritants [25, 26].

Krzysik, D. G. et al. (2002) developed wet wipes or wipe-type products that have improved skin health benefits due to an oil-in-water emulsion composition [25, 26]. The wet wipes or wipe type products of the present invention can be used for baby wipes, adult wipes, hand wipes, face wipes, cosmetic wipes, household wipes, industrial wipes, personal cleansing wipes, cotton balls, cotton-tipped swabs and the like. The natural fat or oil used in the composition may include borage oil, avocado oil, or sunflower oil. The sterol or sterol derivative used in the composition may include soy sterol, avocado sterol, or cholesterol.

7. FUTURE DEVELOPMENTS

The World Avocado Organization (WAO) expects continued strong growth in Europe and the U.K. during 2019, thanks in part to several food trends. As part of its 2019 Avocado forecast, the entity said that the region consumed over 650 million kilos of avocados, adding that Europe remains one of the fastest-growing markets worldwide. With growth expected to continue by at least 15 percent per year in the next five years, the 'Fruit of Life' is in no danger of falling out of favor.

Different varieties of avocado are cultivated worldwide, and Hass and Fuerte are among the most consumed. Avocado industrial processing generates large quantities of agro-industrial byproducts, mainly peel and seed, ranging from 18% to 23% of fruit dry weight [27]. Thus, it is interesting to reuse these byproducts both to reduce their negative impact on the environment and to add value to them, in as much as they are sources of critical phytochemical compounds [19, 20, 28, 29]. Avocado peel and seed have high contents of bioactive phytochemicals such as phenolic acids, condensed tannins, flavonoids, (including procyanidins, flavonols, hydroxybenzoic, and hydroxycinnamic acids) perseitol and sugars such as mannoheptulose [19, 28, 29, 30]. These bioactive compounds have shown various biological activities such as antimicrobial, anti-inflammatory, cardiovascular disease risk reduction, antioxidant, innate immunity strengthening, anticancer, and dermatological activities. Therefore, the transformation of these agro-industrial byproducts, which yield a range of bioactive substances, should be made by green extraction to reduce the environmental impact of the agro-industrial chain. Agro-industrial byproducts are essential sources of natural bioactive compounds that can be employed in several industries. Many food-grade ingredients can be obtained from avocado wastes, particularly premium-grade fats or extracts with high functional power. Moreover, extracts obtained by green extraction can be used as ingredients in the formulation of functional foods. Studies should continue to identify the profiles and phytochemicals available to the business sector, which can also be implemented to valorize the nutritional and functional potential of avocado pulp, seeds, and peels.

REFERENCES

[1] Araújo, R. M., Rodríguez-Jasso, H., Ruiz, M. M., Pintado, C., and Aguilara, N. (2018). Avocado by products: Nutritional and functional properties. *Trends in food science and technology.* 80: 51-60.

[2] Rodríguez-Carpena, J. G., Morcuende, D., Andrade, M. J., Kylli, P., Estevez, M. (2011). Avocado (*Persea americana* Mill.) phenolics, in vitro antioxidant and antimicrobial activities, and inhibition of lipid and protein oxidation in porcine patties. *J. Agric Food Chem,* 59: 5625–5635.

[3] Yahia, E. M. (2009). The contribution of fruit and vegetable consumption to human health. In: *Noureddine B., editor. Postharvest technologies for horticultural crops,* (II):1–25.

[4] Hurtado-Fernández, E., Carrasco-Pancorbo, A., Fernández-Gutiérrez, A. (2011). Profiling LC-DAD-ESI-TOF MS method for the determination of phenolic metabolites from avocado (*Persea americana*). *J. Agric food chem,* 59: 2255–2267.

[5] Figueroa, J. G., Borrás-Linares, I., Lozano-Sánchez, J., Segura-Carretero, A. (2018). Comprehensive characterization of phenolic and other polar compounds in the seed and seed coat of avocado by HPLC-DAD-ESI-QTOF-MS. *Food Res Int,* 105: 752–763.

[6] Figueroa, J. G., Borrás-Linares, I., Lozano-Sánchez, J., Segura-Carretero, A. (2018). Comprehensive identification of bioactive compounds of the avocado peel by liquid chromatography coupled to ultra-high-definition accurate-mass Q-TOF. *Food Chem,* 245: 707–716.

[7] Murakami, Y., Kawata, A., Ito, S., Katayama, T., Fujisawa, S. (2015). Radical-scavenging and anti-inflammatory activity of quercetin and related compounds and their combinations against RAW264.7 cells stimulated with porphyromonas gingivalis fimbriae. Relationships between anti-inflammatory activity and quantum chemical parameters. *In Vivo* (Brooklyn), 29: 701–710.

[8] Wang, W., Bostic, T. R., Gu, L. (2010). Antioxidant capacities, procyanidins and pigments in avocados of different strains and cultivars. *Food Chem,* 122: 1193–1198.

[9] Melo, P. S., Massarioli, A. P., Denny, C., Dos-Santos, L. F., Franchin, M., Pereira, G. E. (2015). Winery byproducts: Extraction optimization, phenolic composition, and cytotoxic evaluation to act as a new source of scavenging of reactive oxygen species. *Food Chem,* 181: 160–169.

[10] Tehranifar, A., Selahvarzi, Y., Kharrazi, M., Bakhsh, V. J. (2011). High potential of agro-industrial by products of pomegranate (Punica granatum L.) as the powerful antifungal and antioxidant substances. *Ind Crops Prod,* 34: 1523–1527.

[11] Hernández-Brenes, C., García-Cruz, M.I., Gutiérrez-Uribe, J.A., Benavides-Lozano, J.A., Rodríguez-Sánchez, D.G. (2012). *Antimicrobial, antibacterial and spore germination inhibiting*

activity from an avocado extract enriched in bioactive compound, WO 2012/042404 A2, April, 5, 2012.

[12] Hernández-Brenes, C., García-Cruz, M.I., Gutiérrez-Uribe, J.A., Benavides-Lozano, J.A., Rodríguez-Sánchez, D.G. (2013). *Antimicrobial, antibacterial and spore germination inhibiting activity from an avocado extract enriched in bioactive compounds,* US 20130216488A1, August, 22, 2013.

[13] Hernández-Brenes, C., García-Cruz, M.I., Gutiérrez-Uribe, J.A., Benavides-Lozano, J.A., Rodríguez-Sánchez, D.G. (2015). *Antimicrobial, antibacterial and spore germination inhibiting activity from an avocado extract enriched in bioactive compounds.* EP 2851062A2, March, 25, 2015.

[14] Piccirilli, A., and Legrand, J. (2004). *Method for producing an avocado leaf extract rich in furanic lipids.* US 20040022882 A1, February 5, 2004.

[15] Frondoza, C., Henderson, T.R., Grzanna, R. (2016). *Orally administrable compositions comprising avocado/soybean unsaponifiable and lipoic acid and methods of administration.* US 9421234 B2. August 23, 2016.

[16] Brovelli, E.A., Vallejos, J.A., Ikonte, Ch., Menon, G.R., Fast, D.J., Chandra, A. (2005). *Compositions containing avocado leaf extract for lowering cholesterol levels.* WO 2005115422 A2 December 8, 2005.

[17] Hsia, H. S., and Fan, D. (2002). *Nutritive composition for cardiovascular health containing fish oil, garlic, rutin, capsaicin, selenium, vitamins, and juice concentrates.* US 006440464 B1, August 27, 2002.

[18] Piccirilli, A., Picardi, N., Msika, P., Paul, F., Bredif, S. (2015). *Use of a compound comprising D-mannoheptulose and/or perseitol for treating and preventing innate immunity modification diseases.* US 009023810B2, May, 5, 2015.

[19] Martinez-Serna, V., Beyer, S. R., Ter Haar, R. H. (2015). *Processes for preparing a carbohydrate extract comprising mannoheptulose*

and compositions comprising same. WO 2015070252 Al, May 14, 2015.

[20] Beyer, S. R., and Ter Haar, R. (2014). *Avocado processing.* US 20140348986 A1 November 27, 2014.

[21] Beyer, S. R., and Ter Haar, R. (2014). *Ingestible compositions containing criollo avocado component.* WO 2014190136 A1, November 27, 2014.

[22] Msika, P. and Legrand, J. (2007). *Process for obtaining a triglyceride-rich refined avocado oil, and oil which can be obtained by means of such a process.* WO 2007057439 Al, May 24, 2007.

[23] Saunois, A., Baudouin, C., Lecleere-Bienfait, S., Garnier, S., Msika, P. (2013). *Avocado flesh and/or skin extract rich in polyphenols and cosmetic, dermatological and nutraceutical compositions comprising same.* US 20130183255 A1 July 18, 2013.

[24] Saunois, A., Baudouin, C., Lecleere-Bienfait, S., Garnier, S., Msika, P. (2018). *Extract of the pulp and/or peel of avocado rich in polyphenols and cosmetic, dermatologic or nutraceutical compositions comprising thereof.* EP2654766B1, October, 10, 2018.

[25] Krzysik, D. G., Lange, B. A., Otts, D. R., Nelson, B. M. (2002). *Wet wipes having skin health benefits. US* 006440437 B1, August 27, 2002.

[26] Krzysik, D. G., Lange, B. A., Otts, D. R., Nelson, B. M. (2001). *Absorbent articles providing skin health benefits.* US 6287581B1, September, 11, 2001.

[27] Wang, W., Bostic, T. R., Gu, L. (2010). Antioxidant capacities, procyanidins and pigments in avocados of different strains and cultivars. *Food Chem,* 122: 1193–1198.

[28] Hurtado-Fernández, E., Carrasco-Pancorbo, A., Fernández-Gutiérrez, A. (2011). Profiling LC-DAD-ESI-TOF MS method for the determination of phenolic metabolites from avocado (*Persea americana*). *J. Agric Food Chem,* 59: 2255–2267.

[29] Figueroa, J. G., Borrás-Linares, I., Lozano-Sánchez, J., Segura-Carretero, A. (2018). Comprehensive characterization of phenolic

and other polar compounds in the seed and seed coat of avocado by HPLC-DAD-ESI-QTOF-MS. *Food Res Int.,* 105: 752–763.

[30] Figueroa, J. G., Borrás-Linares, I., Lozano-Sánchez, J., Segura-Carretero, A. (2018). Comprehensive identification of bioactive compounds of avocado peel by liquid chromatography coupled to ultra-high-definition accurate-mass Q-TOF. *Food Chem.,* 245: 707–716.

In: Avocado: Consumption and Health ISBN: 978-1-53618-204-0
Editors: M. G. Melgar Lalanne et al. © 2020 Nova Science Publishers, Inc.

Chapter 7

CHARACTERIZATION OF GUACAMOLE FORMULATED WITH TWO VARIETIES OF YUCATECAN AVOCADO (*PERSEA AMERICANA*)

V. Toledo López[1], F. Concha-Valdez[2], A. Zarza-García[3], Y. Moguel Ordóñez[4], J. Tepal-Chalé[4], E. E. Martínez-Leo[5] and M. R. Segura-Campos[5,*]

[1]Tecnológico Nacional de México/Instituto Tecnológico de Mérida, Mérida, Yucatán, México
[2]Centro de Investigaciones Regionales "Dr. Hideyo Noguchi", Universidad Autónoma de Yucatán. Mérida, Yucatán, México
[3]Facultad Ciencias de la Salud, Universidad Autónoma del Carmen, Ciudad del Carmen, Campeche
[4]Instituto Nacional de Investigaciones Forestales, Agrícolas y Pecuarias, Mérida, Yucatán, México
[5]Facultad de Ingeniería Química, Universidad Autónoma de Yucatán, Mérida, Yucatán, México

[*] Corresponding Author's Email: maira.segura@correo.uady.mx.

Abstract

The development of new food products requires the study of the interaction of their ingredients with the food matrix, preservation technologies, the nutritional and functional value of the product, as well as acceptance of the same by the consumer. Consumed widely in Mexico, guacamole is formulated from avocados; this can be made from varieties adapted to their cultivation, in order to improve their quality attributes. The objective of this study was to characterize microbiological, chemistry, and sensorially, guacamole formulated with two varieties of avocado: lagunero (L) and double harvest (DH) grown in Yucatán. After the disinfection of the work area, utensils and avocados, the pulp was extracted. Lemon juice (10%) and common salt (2%) was added to each pulp. Subsequently, the mixture was homogenized and placed in sterile flasks. The microbiological analysis was carried out according to NOM-092-SSA1-1994 and NOM-210-SSA1-2014. The proximal composition was determined according to the AOAC (1997). The sensory analysis was carried out through an affective test using a hedonic scale that evaluated attributes of color, taste, texture, smell, overall impression, and purchases intention. The results indicated that the microbiological quality of the guacamoles was within the ranges accepted by the regulations. Aerobic mesophilic bacteria were 190 and 240 CFU/g, respectively. Total and fecal coliforms were <3 MPN/g in both guacamoles; no presence of *Salmonella*. Protein, fat, ash, fiber, and NFE content were 11.59/7.10, 27.1/34.10, 10.6/8.53, 1.9/1.62, 48.74/47.84% for L/DH, respectively. The sensory analysis showed that the evaluated attributes were above the point of indifference with the purchase intention of "maybe I would buy." In conclusion, the characterization of guacamole based in avocado adapted to its cultivation in the Mexican southeast suggests a new traditional product with attributes of improved quality.

Keywords: avocado, guacamole, proximal composition, microbiological analysis, sensory analysis

1. Introduction

Avocado is a tropical fruit increasingly valued by consumers, not only for its particular flavor and texture but for the possible beneficial effects for the health of those who consume it. One of the culturally most accepted by-products is guacamole, which is a preparation made from "crushed" avocado, salt, pepper, and lemon. Scientific and technological development has contributed to the generation of new varieties of avocado and its subproducts, with improved attributes of taste, quality, and perception, which has led to an increase in the number of investigations and the search for potential beneficial effects on consumer health.

Avocado and its by-products (guacamole and oil) are foods with high acceptance and international commercialization, being Mexico, the leading producer of avocado with 1 997 629 tons in 2017, which represented 4.39% of its national agricultural growth domestic product [1]. From 2013 to 2016 the average sales of Mexican avocados to the world was 1.74 billion dollars, with an average annual growth rate of 15.2%, being the main export destinations United States, Japan, Canada, Spain, France, the Netherlands, El Salvador, China, Honduras, and Guatemala, which together accounted for about 98% of the total exports made by Mexico. In 2030, it is estimated that Mexico will have an increase in world demand from 2.84 to 4.24 million tons, with an accumulated growth of 48.98% [2].

The high global avocado consumption has attracted the attention of several research groups, directing their studies towards the benefits towards health and their different forms of consumption, which is why the development and characterization of guacamole are of interest, as well as the knowledge of its nutritional benefits and functional properties.

1.1. Botanical Characteristics of *P. americana* Mill

The avocado is a perennial plant that belongs to the genus Persea of the Lauraceae family, native to Mexico and Central America. The genus Persea that is divided into three different subgenera that include more than 150 species: 1) Persea that only has two species: *P. americana* and *P. schiedeana*; 2) Eriodaphne where about 70 species belong, such as *P. caerulea*, *P. indica*, *P. lingue*, among others; 3) Machilus with around 80 species, such as *P. japonica* and *P. kobu*. The most studied and most important species of the Persea genus is *P. americana*, whose fruit is the commercial avocado (Table 1) [3].

Avocado is the most common English name but is also known as "crocodile pear" and "butter pear." Spanish-speaking people usually call it avocado, "cura" or "cuprandra," although, in Ecuador and Chile, it is called by its Inca name, "palta." The name avocado is derived from the Aztec word Nahuatl, ahuacatl, meaning "testicle." This name refers to the shape of the fruit, considered by the Aztecs as the fruit of fertility [4].

Table 1. Taxonomic hierarchy of *P. americana* Mill

Taxonomy	
Kingdom	Vegetable
Phylum	Fanerogama
Division	Magnoliophyta
Class	*Magnoliopsida*
Order	*Laurales*
Family	*Lauraceae*
Genus	*Persea*
Scientific name	*Persea americana* Mill.
Commun name	Avocado

Table 2. Comparison of the three different avocado horticultural races [8]

Trait	**Race**		
	Mexican	**Guatemalan**	**Antillean**
Climate	Semi-tropical	Subtropical	Tropical
Cold tolerance	High	Intermediate	Low
Salinity Tolerance	Low	Intermediate	High
Size tree	5- 7 m	10- 18 m	6- 8m
Fruit form	Elongated	Round	Variable
Fruit color	Green	Dark green	Reddish
Seed size	Large	Small	Variable
Oil content	Very high	High	Low
Pulp flavor	Sweet	Anise	Soft aroma

The origin of the avocado, according to Williams [5], took place in the upper parts of central and eastern Mexico and the high parts of Guatemala. This same region is known as Mesoamerica and is also considered as the area where domestication was carried out. Three avocado breeds are recognized; Mexican, Guatemalan and Antillean. The botanical classification of these three races has been varied. Some indicate the Mexican race as a separate species; *Persea drymifolia* [6], others like Williams [7] classified the Guatemalan race as *Persea nubigena* var. Guatemalensis. However, the three breeds are currently considered inside the *Persea americana* Mill species (Table 2).

In its natural habitat, the avocado plant reaches heights of 10 to 40 m, with shallow roots, abundant branches, thin and fragile, with flowers from June to September and fruits in October. Mainly, the fruit is the berry of a seed, oval, with a smooth or rough surface, with a comprehensive weight range (120-500g), greenish color, fine skin, with a soft consistency when ripe [2]. The trunk reaches 80 cm in diameter, is erect, with scaly bark and gray. The leaves are entire and straightforward, reddish, and, when ripe, they become smooth, leathery, and of an intense green [9]. The species is widely cultivated in warm zones (200-3000 m.s.). However, it is susceptible to the lack of air in the soil, one of the main factors that affect the production and quality of its fruit (avocado), especially in unfavorable soil conditions, an inadequate relationship between water and air in the zone of the rhizosphere (zone of unique and dynamic interaction between plant roots and soil microorganisms, a product of irrigation management) [10].

1.2. Chemical and Nutritional Characteristics of *P. americana* Mill

Nutritionally, the avocado is a fruit that, unlike many others (high in sugars), has a high lipid content. A piece of approximately 136g provides 227 kcal, 21g of fat, 11.8g of carbohydrates, and 2.7g of protein. Among the major micronutrients present in avocados are potassium (690mg) is found, magnesium (39mg), ascorbic acid (12mg), niacin 2.6mg, and folic acid (1.2mg) [11].

Table 3. Avocado fatty acids

Fatty acid	Content (%)
Palmitic acid	28.21
Palmitoleic acid	5.69
Stearic acid	0.69
Oleic acid	50.95
Linoleic acid	13.87
Linolenic acid	0.58

Mainly, the high lipid composition of the avocado is due to a high content of monounsaturated fatty acids (71%) (Table 3), being mostly oleic acid [12], belonging to the omega nine family and recognized for its cardioprotective properties [13].

Regarding avocado carbohydrate composition, 80% of these represent dietary fiber, which consists of 70% insoluble fiber and 30% soluble fiber [14]. Avocados contain approximately 2g of dietary fiber per 30g [11], which is why its effect on satiety and other benefits on digestive health has also been studied [15].

1.3. Potential Health Benefits of *P. americana* Mill

The high content of monounsaturated fatty acids, mainly oleic acid (omega 9) and secondary metabolites such as lutein, carotenes, zeaxanthin, phenolic compounds and phytosterols such as β-sitosterol, campesterol, and stigmasterol, has been the main reason to study the functional benefits of avocado consumption in the prevention and treatment of chronic-metabolic diseases, such as obesity, diabetes, hypertension and even metabolic syndrome [16].

Hypoglycaemic effects have been reported in rats, after oral administration of ethanol extracts of avocado at a dose of 300 mg/kg for four weeks, with a decrease also in levels of glycated hemoglobin, urea and serum creatinine [17]. Likewise, consumption for eight weeks exhibited a decrease in body weight, an increase in muscle mass, and reduction of hepatic transaminase levels [18]. The consumption of aqueous avocado extracts at doses of 300 and 600 mg/kg, showed a significant reduction of blood glucose levels in diabetic rats induced with alloxan, with a protective effect on the beta cells of the pancreas [19] while, at doses of 100 and 200 mg/kg for two months, showed a significant reduction of total cholesterol, triglycerides, and LDL in rabbits [20].

Ethanolic extracts of avocado at doses of 450 and 900 mg/kg have shown a significant decrease in triglycerides, total cholesterol, and low-density lipoprotein (LDL) in diabetic rats induced with alloxan [21]. It is also known the hepatoprotective effect of avocado fruit, in rats with liver damage induced with carbon tetrachloride, where after consumption for four weeks of 10 and 15% of avocado nuts, decreased serum transaminase levels were observed, total and direct bilirubin, malondialdehyde (MDA) and a significant increase in the enzymatic activity of superoxide dismutase (SOD), glutathione peroxidase (Gpx) and catalase (CAT) [22].

In the treatment of obesity, consumption of avocado has shown an effect on the bodyweight of obese rats under various mechanisms of action, such as increased expression of adiponectin and PPAR-γ mRNA [23]; increased catabolism of lipids accumulated in adipose tissue [24]; increased expression of FASN, LPL, and leptin mRNA in subcutaneous and visceral adipose tissue [25], as well as decreased oxidative stress.

Clinical evidence highlights the functional potential of avocado consumption for the treatment of chronic metabolic diseases. So, in people with diabetes mellitus, consumption of a half piece of Hass avocado significantly reduced blood insulin, and glucagon-1-like peptide levels. This hypoglycemic effect is attributed to the content of D-mannoheptulose present in this fruit [26].

In people with overweight, avocado consumption showed an increase in postprandial insulin levels, favoring satiety [15] while, in people with obesity, the consumption of one piece of avocado for six weeks decreased body weight, BMI, percentage of fat and serum lipids [27]. Recently, the consumption of avocado juice for 14 days reported a decrease in LDL cholesterol levels in people with dyslipidemia [28].

2. METHODS

2.1. Guacamole's Elaboration

The present study used as raw material two varieties of avocado from the Yucatan Peninsula Lagunero (L) and Double Harvest (DH)). The work area and utensils were disinfected, as well as the avocados. Then, the pulp was extracted, each in different containers, making a total of approximately 500 g for DH and 400 g for L. Each one was added 10% lemon juice and 2% common salt. Once the guacamoles were obtained, they were analyzed microbiologically, proximally, and sensorially to determine the quality of the guacamole and to determine possible differences between them.

2.2. Microbiological Analysis

The guacamole samples were collocated in sterile glass bottles of 120 mL and remained in refrigeration until their analysis.

2.2.1. Analysis of Viable Microorganisms (Aerobic Mesophiles)

The procedure was carried out according to NOM-092-SSA1-1994 [29], using a 1:10 dilution of each sample with phosphate-buffered saline solution 1x (PBS) and 15 mL of standard agar (DIFCO) in sterile Petri dishes. Subsequently, they were mixed by six movements from right to left, six in the clockwise direction, sic in the opposite direction, and six from back to front, on a smooth horizontal surface until complete incorporation of the inoculum in the medium was achieved. A petri dish without inoculum was included as a control. The dishes were incubated in the inverted position (the lid down) at 35 ± 2°C for 48 ± 2h. Next, the colonies were counted and reported as colony-forming units per gram (CFU/g).

2.2.2. Total Coliforms and Fecal

This analysis was carried out in Appendix H of NOM-210-SSA1-2014 [30].

2.2.3. Presumptive Test

25g of products were suspended in 225 mL of PBS. Decimal dilutions were prepared with the PBS. Lauryl tryptose broth was used as an enrichment medium; once inoculated, the tubes were incubated at 35 ± 0.5°C. The tubes were examined at 24 h, and it was observed if there was gas formation.

2.2.4. Confirmatory Test for Total Coliforms

Two or three samples, approximately every positive tube obtained during the presumptive test, were transferred to tubes containing bright green bile broth in a Durham bell. The tubes were gently shaken for homogenization and incubated at 35 ± 2°C for 24 to 48 h. The tubes where growth and gas production were observed after an incubation period of 24 to 48 h were recorded as positive. The Most Likely Number tables (MLN) were used to know the most probable number of total coliform organisms/100 mL.

2.2.5. Confirmatory Test for Fecal Coliforms

It was the same process as for total coliforms, with the difference that the *E. coli* broth (ECB) was used as culture medium, also in a Durham bell, and the incubation temperature was 44.5 ± 0.1°C.

2.2.6. Detection of Salmonella ssp.

It was carried out by the procedure described in Appendix A Normative of the same NOM-210-SSA1-2014 [30], where 25 g of the sample was diluted with 225 mL of buffered peptone water. This method consists of 4 stages: a pre-enrichment, a selective enrichment, selective isolation, and biochemical identification, and serological confirmation of the microorganisms. Finally, the result was reported as *Salmonella* spp in 25 g of sample, present or absent.

2.3. Chemical Analysis

The chemical composition of the products was determined according to the official procedures described by AOAC [31], which includes the following analyzes:

a) Moisture (method 925.09), was determined by weight loss after drying the sample at 105°C for four h.
b) Raw protein (method 954.01), was determined by the Kjeldahl method by acid digestion (H_2SO_4) of the sample and alkaline distillation (NaOH); using 6.25 as a factor for the conversion of nitrogen to protein.
c) Raw fat (method 920.39), it was quantified after the extraction of the sample during one h with hexane in Soxleth equipment.
d) Raw fiber (method 962.09), was calculated after acid and alkaline digestion with a Fibertec system (Tecator Sweden).
e) Ash (method 923.03), was calculated as the remaining weight after calcining the sample in a muffle at 550°C for four h.
f) Total Carbohydrates (Nitrogen Free Extract-NFE), was estimated by 100% difference of the rest of the constituents.

2.4. Sensory Analysis

This analysis was carried out employing an affective test, using a hedonic scale, structured with seven descriptive points in which the characteristic of liking was pointed out. This scale had an indicator of the midpoint (4 points) to provide each consumer judge the facility to find a point of indifference to the product in question [32]. Higher points of 4 indicated levels of liking; I like a little (5), I like it (6), and I like it a lot (7); points less than four referred to levels of displeasure, I dislike a little (3), I dislike (2), and I dislike a lot (1). Thus, 80 untrained judges evaluated the attributes of color, taste, texture, smell, and overall impression. Likewise, the purchase intention of both products was evaluated using a 5 point scale, where the points above 3 indicated purchase intentions: probably buy (4), surely buy (5) and, the points less than three intentions of not purchase: probably would not buy (2), probably would not buy (5).

2.5. Statistic Analysis

All the results were processed by descriptive statistics, using measures of central tendency (average) and dispersion (standard deviation). The data obtained from the chemical and sensory analysis were evaluated using one-way analysis and comparison of means (LSD method) to establish differences between treatments, with a confidence level of 95% ($p < 0.05$) [33].

3. Results and Discussion

3.1. Microbiological Analysis

The results showed that the microbiological quality of the guacamoles prepared with the care and hygiene applied, was acceptable; that is, low amounts of aerobic mesophilic bacteria were obtained (190 and 240 CFU/g for DH and L, respectively), total and fecal coliforms (<3 MPN/g), as well as the absence of *Salmonella* in both guacamoles.

3.2. Chemical Analysis

The results are presented in Table 4.

Table 4. Chemical composition of guacamole made with two varieties of Yucatecan avocados

Parameter (%)	Lagunero avocado Average ± s.d.	Double Harvest avocado Average ± s.d.
Moisture	88.96 ± 0.16	85.22 ± 0.73
Protein	11.59 ± 0.34	7.10 ± 0.76
Fat	27.17 ± 0.70	34.91 ± 0.36
Ashes	10.60 ± 0.16	8.53 ± 0.74
Fiber	1.90 ± 0.12	1.62 ± 0.08
NFE	48.74 ± 0.96	47.84 ± 0.58

The moisture found in the guacamoles made with Yucatecan avocados was higher than those reported for commercial avocados, this due to the addition of 10% lemon juice in its formulation. Morillas and Delgado [34] published for avocados, Haas variety of Peruvian origin, percentages of humidity that were from 68.0 to 78.8%. However, the highest moisture content has been reported in Creole-type avocados [35] with local avocados from Colombia, with percentages ranging from 75.85 to 81.24%. This tendency to present a higher moisture content in creole avocados could have been another factor that influenced the high moisture content in both guacamoles. The raw protein content of guacamole made with avocado L was higher than that made with DH. The last one was found within the values reported for creole avocados with values of 5.19 to 6.06%. The content of raw fat was higher in guacamole made with avocado DH (34.91%), which is an advantage since the oil contained in the avocado is an essential source of fatty acids beneficial to health. It has been reported lipid values in avocado paste ranging from 21 to 33%, which contain a high percentage of unsaturated fatty acids (74-83%) and vary depending on the variety and stage of maturity of the avocado [36]. The ash content found resulted in higher in both guacamoles than those reported [35] mainly for the addition of salt to the formulation. The crude fiber values found in the guacamoles were within the range of 1.8 to 9.11% reported for avocados [34], and the content of NFE that represents the content of sugars and polysaccharides of vegetable reserves represented approximately 50% of the dry matter in the guacamoles.

3.3. Sensory Analysis

This analysis showed the level of liking of the guacamoles made with avocado L and DH both products registered an average of ratings above the point of indifference, that is, values greater than 4 of the hedonic scale that they inferred "I do not like or dislike" (Figure 1).

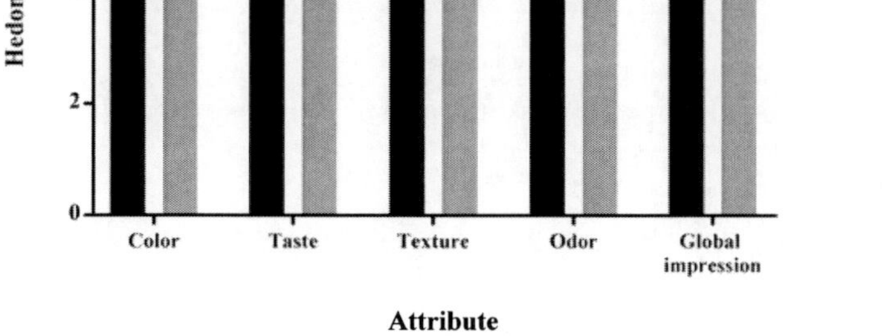

[a]Equal letters mean statistical equality ($p > 0.05$).

Figure 1. Average of grades emitted in the sensory analysis of guacamole elaborated with avocado Lagunero (L) and Double Harvest (DH).

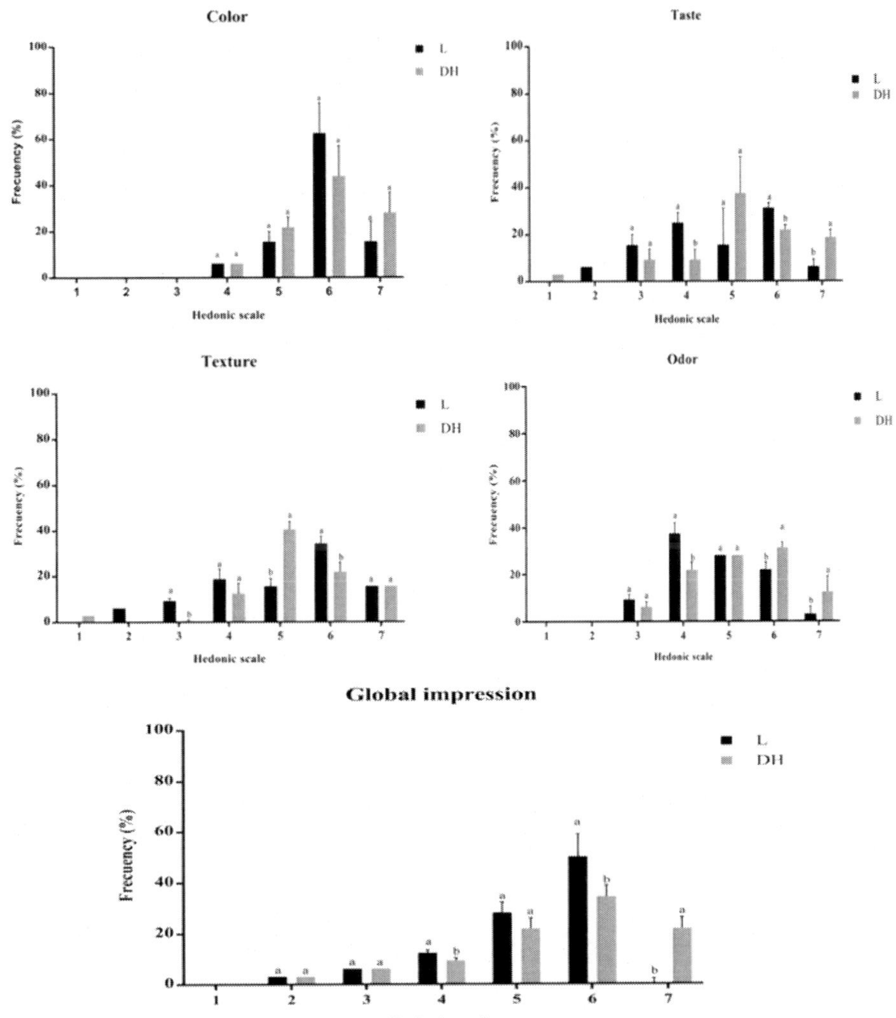

[a-b]Different Letters at the same level as the hedonic scale mean statistically significant difference ($p < 0.05$).

Figure 2. Percentage of frequency achieved in the attributes evaluated in the guacamole elaborated with Lagunero (L) and Double Harvest (DH).

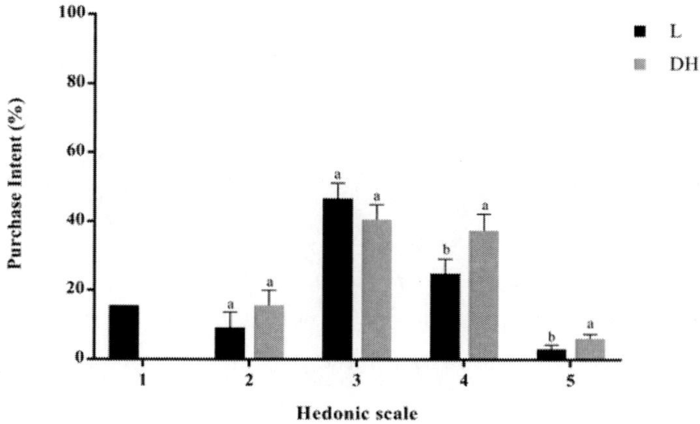

[a-b]Different Letters at the same level as the hedonic scale mean statistically significant difference ($p < 0.05$).

Figure 3. The intention of purchase obtained in the sensory analysis of guacamole elaborated with avocado Lagunero (L) and Double Harvest (DH).

The percentage of registered frequencies, for each level of the hedonic scale and, in all evaluated attributes corroborates that, the untrained judges of this study expressed an appreciation for both guacamoles, judging their color, taste, texture, smell, and impression global above the point of indifference (level 4). However, it was the guacamole made with DH avocado that recorded, except for the texture attribute, the highest percentage of frequency in level 7 of the hedonic scale ("I like it a lot") for all the evaluated attributes being statistically different ($p < 0.05$) to the percentage recorded for guacamole made with avocado L (Figure 2). The result of purchase intention obtained in the sensory analysis showed that the judges would have higher intention to buy the guacamole made with avocado DH, since it registered higher percentages of said intention in levels 4 (probably would buy) and 5 (probably would buy) the scale used, versus the guacamole elaborate with avocado L (Figure 3).

Since the sensory analysis registered the pleasure of the products made with both avocados, it is interesting to continue investigating the physicochemical properties and the functional potential of the guacamoles, thus promoting the development of new products and the production not only local but also national and international of the double harvest variety.

CONCLUSION

The microbiological study and the chemical and sensory analysis of guacamole made with avocado L and DH, highlight the quality of the products, as well as the nutritional value and the level of pleasure among consumer judges. It is interesting to continue studying the biological potential of guacamoles, associated with the lipid profile of avocado DH, thus contributing to the development of new products, of interest not only for its nutritional value but also for the health benefits that could derive from its consumption.

REFERENCES

[1] Servicio de Información Agroalimentaria y Pesquera (SIAP). *Atlas agroalimentario 2012-2018.* México. Disponible en: https://nube. siap.gob.mx/gobmx_publicaciones_siap/pag/2018/ Atlas-Agroalimentario-2018.

[2] Secretaría de Agricultura, Ganadería, Desarrollo Rural, Pesca y Alimentación (SAGARPA) [Fishing and Food (SAGARPA)] (2017). *Planeación Agrícola Nacional 2017-2030.* Disponible en: https://www.gob.mx/cms/uploads/ attachment/file/257067/Potencial-Aguacate.pdf.

[3] Gómez, M., Escalera, D., Rojas, P., Valeriano, M., López, A., Muruchi, A. (2017). *Persea americana* Mill. (Palta) seed benefits. *Revista de Investigación e Información en* Salud. 12(30): 38-43.

[4] Brom, R. E., Carvalho, C. F. (1966). *El aguacate* [*Avocado*]. Ed. Juan Lozoya Dávila: 122.

[5] Williams, L. O. (1977a). The avocados, a synopsis of the genus *Persea, sung. Persea. Economic Botany* 31: 315-320.

[6] Kopp, L. E. (1966). A taxonomic revision of the genus Persea in the Western Hemisphere (*Perseae-Lauraceae*). *Memoirs of the New York Botanical Garden*. 14(1): 1-120.

[7] Williams, L. O. (1977b). The botany of the avocado and its relatives. *Proc. 1st International Tropical fruit Short Course, The Avocado. The* University of Florida, Gainesville, Florida. USA: 9-15.

[8] Paull, RE., Duarte, O. (2010). *Avocado in Tropical fruits* (crop production science in horticulture). R. E. Paull and O. Duarte (Eds.). CABI Publishing, Wallingford, United Kingdom:153-184.

[9] *Ministerio de Agricultura Ganadería y Alimentación*. (2000). El cultivo del aguacate en Guatemala.

[10] InfoAgro. (2014). *El cultivo del aguacate de Fruticultura Subtropical* (En línea) [*The cultivation of the avocado of Subtropical Fruit Growing* (Online)]. Disponible en: http://www.infoagro.com/frutas/frutas_tropicales/aguacate.htm [Acceso: 12 marzo de 2019].

[11] USDA (US Department of Agriculture). (2011). Avocado Composition. Nutrient Data Laboratory. *USDA National Nutrient Database for Standard Reference, Release 24*. Washington, DC: US Department of Agriculture.

[12] Lu, Q. Y., Zhang, Y., Wang, Y., Wang, D., Lee, R. P., Gao, K., Byrns, R., Heber, D. (2009). California Hass avocado: profiling of carotenoids, tocopherol, fatty acid, and fat content during maturation and from different growing areas. *Journal of Agricultural and Food Chemistry*. 57(21):10408-13.

[13] Shores, D., Binion, D., Baker, P. (2011). New Insights into the Role of Fatty Acids in the Pathogenesis and Resolution of Inflammatory Bowel Disease. *Inflammatory Bowel Diseases*. 17(10): 2192–2204.

[14] Marlett, J. S., Cheung, T. F. (1997). Database and quick methods of assessing typical dietary fiber intakes using data for 228 commonly consumed foods. *Journal of the American Dietetic Association*. 97:1139,–1148.

[15] Wien, M., Haddad, E., Oda, K., Sabaté, J. (2013). A randomized 3x3 crossover study to evaluate the effect of Hass avocado intake on post-ingestive satiety, glucose and insulin levels, and subsequent energy intake in overweight adults. *Nutrition Journal*. 12: 1.

[16] Dabas, D., M. Shegog, R., R. Ziegler, G., D. Lambert, J. (2013). Avocado (*Persea americana*) Seed as a Source of Bioactive

Phytochemicals. *Current Pharmaceutical Design*. 19(34): 6133-6140.

[17] Rao, U. S. M., Adinew, B. (2011). Remnant β-cell-stimulative and antioxidative effects of *Persea americana* fruit extract studied in rats introduced into streptozotocin-induced hyperglycaemic state. *African Journal of Traditional, Complementary and Alternative Medicine*. 8: 210–217.

[18] Thenmozhi, A., Shanmugasundaram, C., Rao, UM. (2012). Biochemical evaluation of anti-diabetic phytomolecule through bioactivity guided solvent fractionation and subfractionation from hydromethanolic (2:3) extract of Alligator pear Fruit in streptozotocin-induced diabetic rats. *Journal of Applied Pharmaceutical Science*. 2: 61.

[19] Edem, D., Ekanem, I., Ebong, P. (2009). Effect of aqueous extracts of alligator pear seed (*Persea americana* Mill) on blood glucose and histopathology of pancreas in alloxan-induced diabetic rats. *Pakistan Journal of Pharmaceutical Science* 22: 272–276.

[20] Nwaoguikpe, R., Braide, W. (2011). The effect of aqueous seed extract of *Persea americana* (avocado pear) on serum lipid and cholesterol levels in rabbits. *African Journal of Pharmacy and Pharmacology*. 1: 23–29.

[21] Edem, D. (2010). Antihyperlipidemic effects of ethanolic extracts of alligator pear seed (*Persea americana* Mill) in alloxan-induced diabetic rats. *Pharmacologyonline* 1: 901–908.

[22] Mahmoed, M. Y., Rezq, A. A. (2013). Hepatoprotective effect of avocado fruits against carbon tetrachloride-induced liver damage in male rats. *World Applied Science Journal*. 21: 1445–1452.

[23] Padmanabhan, M., Arumugam, G. (2014). Effect of *Persea americana* (avocado) fruit extract on the level of expression of adiponectin and PPAR-γ in rats subjected to experimental hyperlipidemia and obesity. *Journal of Complementary and Integrative Medicine*. 11:107–119.

[24] Brai, B. I., Odetola, A. A., Akindele, S. K., Fesobi, T. W., Agomo, P. U. (2012). Evaluation of antiperoxidative and antioxidant properties

of aqueous and methanolic leaf extracts of *Persea americana* Mill. in rats fed high lipid diet. *Canadian Journal of Pure and Applied Science* 6: 2079–2088.

[25] Monika, P., Geetha, A. (2015). The modulating effect of *Persea americana* fruit extract on the level of expression of fatty acid synthase complex, lipoprotein lipase, fibroblast growth factor-21, and leptin – A biochemical study in rats subjected to experimental hyperlipidemia and obesity. *Phytomedicine*. 22: 939–945.

[26] Sabaté, J., Wien, M., Haddad, E. (2015). Post-ingestive effects of avocados in meals on satiety and gastric hormone blood levels. *Human Health Nutrition*. 459–461.

[27] Pieterse Z., Jerling J. C., Oosthuizen W., et al. 2005. Substitution of high monounsaturated fatty acid avocado for mixed dietary fats during an energy-restricted diet: effects on weight loss, serum lipids, fibrinogen, and vascular function. *Nutrition* 21: 67–75.

[28] Rahman, S. (2019). Effect of Avocados to LDL Cholesterol as a Preventive Risk of Atherosclerosis. *International Journal of Multidisciplinary and Current Research*. 7: 4-7.

[29] Diario Oficial, 1995. NORMA Oficial Mexicana NOM-092-SSA1-1994, Bienes y servicios. *Método para la cuenta de bacterias aerobias en placa*.

[30] Diario Oficial, 2015. NORMA Oficial Mexicana NOM-210-SSA1-2014, Productos y servicios. Métodos de prueba microbiológicos. Determinación de microorganismos indicadores [Official Gazette, 2015. Official Mexican STANDARD NOM-210-SSA1-2014, Products and services. Microbiological test methods. Determination of indicator microorganisms]. *Determinación de microorganismos patógenos*.

[31] A. O. A. C. (1997). *Official Methods of Analysis. Association of Official Analytical Chemists*, 15th ed.; William Horwitz Editor: Washington.

[32] Anzalduá, A. (1994). *La evaluación sensorial de los alimentos en la teoría y la práctica* [*Sensory evaluation of food in theory and practice*]. Ed. Acribia. Zaragoza, España.

[33] Montgomery, D. (2003). *Diseño y análisis de experimentos* [*Design and analysis of experiments*]. Editorial Limusa S. A. de C. V. México, D. F. pp 21-125.

[34] Morillas-Ruiz JM, Delgado-Alarcón JM. (2012). Análisis nutricional de alimentos vegetales con diferentes orígenes: Evaluación de capacidad antioxidante y compuestos fenólicos totales [Nutritional analysis of plant foods with different origins: Evaluation of antioxidant capacity and total phenolic compounds]. *Nutrición Clínica y Dietética Hospitalaria*, 32(2): 8-20.

[35] Ceballos AM, Montoya, S. (2013). Evaluación química de la fibra en semilla, pulpa y cáscara de tres variedades de aguacate [Chemical evaluation of the fiber in seed, pulp and peel of three varieties of avocado]. In: *Biotecnología en el Sector Agropecuario y Agroindustrial*. Vol 11, No. 1 pp: 103-112.

[36] Ortiz MA, Dorantes AL., Galindez MJ., Cardenas SE. (2004). Effect of a novel oil extraction method on avocado (*Persea americana* Mill) pulp microstructure. *Plant Foods for Human Nutrition*. 59(1):11-14.

In: Avocado: Consumption and Health ISBN: 978-1-53618-204-0
Editors: M. G. Melgar Lalanne et al. © 2020 Nova Science Publishers, Inc.

Chapter 8

STUDIES ON THE APPLICATION OF ELECTRIC FIELDS TECHNOLOGY FOR ENHANCING QUALITY PARAMETERS OF AVOCADO

María Reyna Robles López[1,*], José Hugo Castorena García[2], José Alberto Ariza Ortega[3], María Guadalupe Méndez Ramos[4] and Raúl René Robles de la Torre[1]

[1]Centro de investigacion en Biotecnología Aplicada, Instituto Politécnico Nacional, Tlaxcala, México
[2]Instituto Tecnologico del Altiplano de Tlaxcala, San Diego Xocoyucan, Tlaxcala, México
[3]Interdisciplinary Research Center, Autonomous University of the State of Hidalgo, Pachuca, México
[4]Universidad Tecnológica de la Costa Grande de Guerrero, Petatlán, Guerrero, México

[*] Corresponding Author's Email: mreynarobles@yahoo.com.

ABSTRACT

Food preservation techniques are usually based either on millenary techniques: such as the use of heat, salt, and smoke, along with century-old techniques, such as pasteurization, cold, the use of chemical additives, and pH modification. However, since the second half of the 20th-century, researchers from all over the world began to develop new preservation techniques, and terms like "intermediate moisture content" and "combined methods" were currently in that way. Subsequently, the concept of emerging technologies was finally adopted, the main features of this group of technologies are they use less energy or use it more efficiently. Techniques belonging to this group are high hydrostatic pressures, microwaves, pulsed electric fields, high light pulses, ultrasound, irradiation, and electric fields, among others. This chapter describes the development of an electric field generator equipment and its potential application for conservation of avocado paste, its oils, the whole fruit, in addition to different studies conducted to evaluate its effectiveness in some other agro-products.

The first aim of this research was to demonstrate the effect of electric fields (EF) on the activity of polyphenol oxidase enzyme (PFO), responsible for enzymatic darkening in the avocado paste, achieving an inactivation of 65%. The second was on the search of adverse effects of the EF on the avocado components, such as fatty acids and the vitamins present on avocado paste. After applying the EF on avocado paste, the formation of trans-Fatty acids was not observed. Respect the vitamins that were analyzed (A, D, C, E, and K). The use of EF did not reduce the content of vitamins when they are in the avocado matrix, except for vitamin C, which decreased to 97%. Electric fields technology is an emerging methodology in which energy consumption is minimal. There is no generation of heat, and is considered as a clean production technique and involve less damage to the processed foods. To sum up, this technology can potentially be applied to different types of food products to increase their quality characteristics as well as the shelf life of fruits as avocado.

Keywords: electric field technology, food conservation

1. INTRODUCTION

Mexico is the leading producer, exporter, and consumer of avocado worldwide. In recent years its consumption has increased in 26 countries, including members of NAFTA, TPP, as well as China, among others, all together consume 1.7 million tons. In Mexico, the value of avocado exports could be of the order of 4.655 million dollars, by the 2030 year, (with the price of 2016) (National Agricultural Planning 2016). In Mexico, the national production of avocado in 2018 reached one million 960 thousand tons, with Michoacán being the leader in production. In the last Super Bowl, 35 tons of avocado were consumed in the United States, (Mexbest Safety and Quality Taste 2019). The avocado fruit, due to its economic value, has been described as "green gold," because of its characteristics of flavor, texture and nutritional value, from it, stand out the quality of its lipids and its content of carotenoids and vitamins A, C, and E, the DL-α-tocopherol, all of them act as antioxidant agents.

All the above makes avocado consumption associated with a reduction in the risk of suffering from cardiovascular diseases and several types of cancer, (Tovar 2003), one of the reasons for its importance in health is the number of lipids it contains (60- to 80% dry mass), and their quality. Avocado lipids include saturated fatty acids, palmitic (16: 0) and stearic (18: 0), monounsaturated fatty acids, oleic (18: 1) and palmitic acidic (16: 1), and polyunsaturated fatty acids: linoleic (18: 2) and linolenic (18: 3). Of all the lipids, oleic acid predominates, representing about 60% of total lipids (Mazliak, 1965a, Kikuta and Erickson, 1968, Luza et al., 1990). In general, the combination of fatty acids in the triacylglycerol fraction determines the physical and nutritional characteristics of lipids (Requejo-Tapia et al., 1999).

Because of its moisture content and its excellent nutritional value, avocado is a highly perishable product, apart from microbial attack, one of the factors that most affect avocado is the rapid darkening of its pulp in contact with air. The action of the native enzyme, polyphenol oxidase, produces this darkening, hence the primary form of consumption is as fresh fruit, although there has also been the idea of industrializing it to prolong shelf life and facilitate its export. However, in the world avocado market, only products such as guacamole and avocado pulp are found, which must remain in refrigeration even as a frozen avocado paste. It is also possible to find other products such as avocado oil and by-products from its bone. The

Lyophilized avocado paste does not require refrigeration; however, the sale prices are high.

On the other hand, due to the growing demand for minimally processed foods or with a minimum of preservatives, chemical methods have been sought that inhibit the enzymatic activity of polyphenol oxidase (PPO) which causes the darkening of the avocado pulp, without damaging the sensory and nutritional characteristics of the fresh product. In order to avoid the oxidative rancidity of avocado puree-based products, as shown in Soliva-Fortuny, et al., (2002) article, sorbic acid was added as an antioxidant, the formation of peroxides was inhibited for periods of at least 12 weeks, with little effect due to the presence of sorbic acid and the availability of oxygen. On the other hand, the absence of tocopherol in the product formulation preceded the formation of secondary oxidative metabolites and the rapid saturation of the fatty acid chains. Therefore, sorbic acid has a moderate pro-oxidative effect ($p < 0.05$).

By contrast, in the area of food preservation, since the second half of the 20th century, the combination of preservation methods was sought to avoid the excessive heat of sterilization, for example, pasteurization was combined with refrigeration to extend the shelf life of the milk. The combined methods followed the methods of minimal processing with the same idea of damaging as little as possible the product, subsequently followed the proposal to develop new techniques of food preservation. In that way, in the last decade of the twentieth century, the term emerging technologies or non-thermal technologies was coined, which seek to conserve food, but avoiding the addition of chemical preservatives and without damaging the natural physical and nutritional characteristics thereof by excessive use of heat.

Emerging conservation technologies are characterized by using less energy or using it more efficiently, and its goal is to reduce the use of heat as the primary conservation factor. It is sought that the temperature increases are minimal or nil, having as final products safe food but with sensory and nutritional characteristics close to natural products, as described by Ume Roobab (2018), which highlights the importance of non-thermal technologies and the achievements obtained over time, about the quality of various foods. The bactericide effect of several technologies and the inhibition of enzymatic activity were observed (Ume et al., 2018). The group of emerging technologies consists of high hydrostatic pressures (HHP), microwaves (MO), high-intensity light pulses (HILF), the pulsed electric field (PEF), the electric field (EF), ohmic heating (OH), ultrasound (US), among others. However, some of them are not new, such as high pressures, or irradiation, which are currently back for implementation on a larger industrial scale for the efficiency of energy use.

The interaction of the matter with electrical energy is not recent. At the beginning of the 20th century, the electric current was used to sterilize liquid foods. However, this practice became disused due to the cost of electric power and the appearance of other novel techniques. In the sixties, Doevenspeak (1961), rescues this technology and patents a device to give discrete discharges of electric current, that is how the technology of the Pulsed Electric Field (PEF) was born (Doevenspeck 1961). Subsequently, Sale and Hamilton (1967) were the first to carry out systematic studies on the effect of the pulsed electric field on bacterial cultures, finding that the PEF had effects on the membrane permeability of microorganisms and that this effect could be reversible to a certain point, after which the bacteria died by the disintegration of the cells. As a result, from these studies and other, the electroporation technique was developed, which is currently a conventional technique in molecular biology laboratories around the world, which is used to promote the formation of pores in the membranes and thus exchange material with the interior of bacterial cells.

After this technique, the electric field technique is derived, with a very particular and essential characteristic feature, because in this technique, there are no electric current discharges. The material to be treated is placed in the middle of two electrodes, and, because in this technology there is no contact between the food and electrodes, and there is no electric current passing through the liquid food, the food material is then exposed only to the effect of electrical charges, that are accumulated in the electrodes forming the "electric field."

Coulomb (1784) introduced the concept of the electric field, and later it was taken into the electromagnetic theory to describe the forces that exist between electric charges. Two charges of different signs tend to come together, while two charges with the same sign tend to repel one another.

When for a physic reason two charges are separated, the space between them creates a field; if a third charge falls in the center of that space, it could be attracted to the particle that has the opposite sign, that attraction might have a magnitude and a direction depending on the sign of its charge. The intensity of the attraction or repulsion is the product of the electric charges themselves and inversely of the distance between them. (Sears, Zemansky 2012).

In this chapter, we describe the advances of the use of this non-thermal technology known as the electric field (EF) in fresh avocado to reduce the enzymatic darkening without altering the characteristics of the avocado and without using chemical preservatives that can alter its flavor or smell.

The CIBA-IPN working group set itself the task of having a piece of equipment with the characteristics to achieve the objectives above. In this way, the group designed and built an electric field generator (EF), whose main foundation is the generation of an electric field produced between two electrodes, connected to a control system with direct energy, but the treated sample is not in direct contact with it. As in the case of the pulsed electric field (PEF), or the case of Ohmic heating (OH). When there is no contact between the sample and the electrodes, there is no increase in the temperature generated. The electrodes are placed horizontally in parallel, inside an acrylic chamber, where one of them works as a base to place the samples, the electrodes, wired to the power source, where the intensity of the generated field can be regulated between them, therefore the *field* depends on the applied voltage and is inversely proportional to the distance between the electrodes. The system also allows modifying the frequency of change in the polarity of the direct current, which can be zero (direct current) or with a value positive in frequency.

Table 1 summarizes the differences between these two systems, the traditional PEF and the electric field alone (EF), developed at the CIBA-IPN. Although initially, these two technologies were confused, there is currently a clear distinction between them; in some cases, the high voltage electric field has been called *cold plasma* (Rohit, Chaitanya and Uday S. 2014).

Table 1. Differences between PEF and EF equipment

Characteristics	PEF	EF
Treatment for liquids	Yes	Yes
Treatment for solids	Some	Yes
Continuous process	Yes	No
Increases in temperature	Yes	No
Contact electrode shows	Yes	No
Allows changes in electrode separation	No	Yes
Requires cooling system	Yes	No
It affects the conductivity of the food	Yes	No
There are electric shocks	Yes	No
Cost of the equipment	30-40,000 USD	12,000 USD

Figure 1 shows a diagram with the arrangement of the four main components of the electric field generator system, in Figure 2, the photographs present the four main components of the same. The equipment designed by the CIBA-IPN is easy to handle; it adjusts the distance between the electrodes with an endless screw and allows to manage a variable voltage with the voltage regulator that gives to the equipment output of 0 to 70 kV from a power source (Dielectric test set, Megger ®). All these features give the system a whole range of possibilities in the operating conditions. Also, it is possible to change the frequency of the output voltage, which can range from 5 Hz to 25 MHz and, it is also possible to choose the shape of the wave, which can be square or senoidal, this is chosen in the function generator (B & K Precision 4017A), the wave can optionally be monitored, in an oscilloscope (Tektronix TDS 1001B). It is important to emphasize that this versatility in the elements provides a wide range of possibilities for experimentation with the electric field, emphasizing that the sample can be only a few milligrams to fruits or full products of several tens of grams. Studies in this type of equipment allow the development of necessary studies for the engineering and scaling of these systems. Another important aspect is that these studies can also lay the foundations for the development of this technology to make it work continuously.

The research carried out with this equipment, related to avocado, starts with the aim of reducing the activity of Polyphenol-Oxidase (PPO), and once this possibility crystallized, new unknowns arose, such as: what happens to the fatty acids contained in the pulp after the EF treatment? What happens to the ratio of the cis-trans isomers after the electric field exposure? What happens with the vitamins present in the avocado? This project was carried out to answer the previous questions, and they are described in the following part.

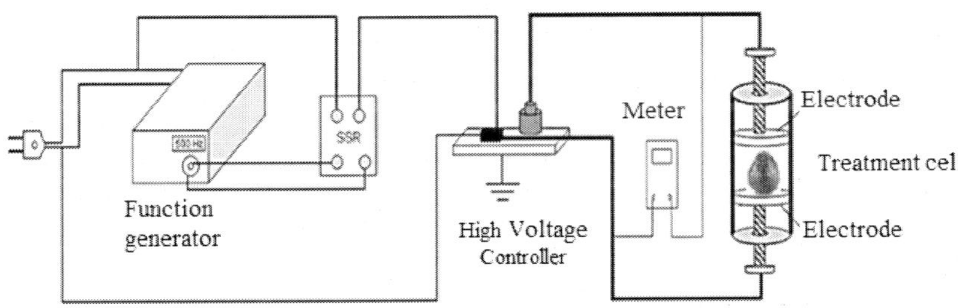

Figure 1. Schematic of the electric field generator CIBA-IPN.

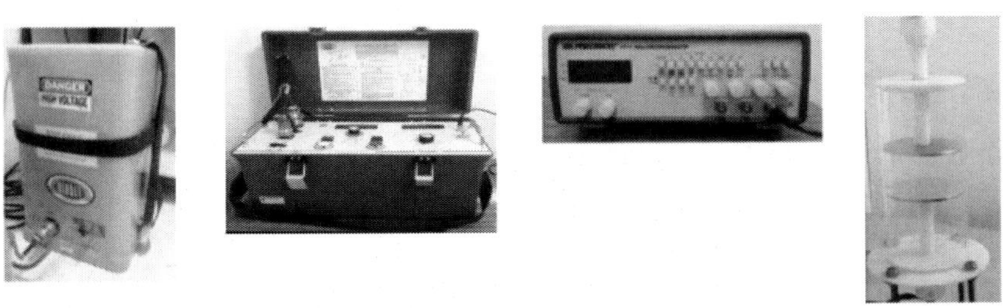

Large power source | Voltage control board | Frequency modulator | Treatment chamber

Figure 2. Photographs of the components of the EF generating equipment.

2. Study of the Effect of the Electric Field on Avocado Components

2.1. Application of the Electric Field on the Activity of the PPO Enzyme for Conservation Purposes

2.1.1. Introduction

The equipment to generate the electrical field, described in the last section, designed and built with the primary objective of evaluating if the electric field generated, could have any effect and of what type, on the avocado paste, particularly on the activity of the enzyme poly-phenol-oxidase (PPO). This enzyme causes the enzymatic darkening of various fruits, such as apples, peaches, and avocados, among others. When the fruit has a blow, fracture, or a cut, the enzyme has contact with the air, and the series of reactions that give rise to the said enzymatic darkening begins. From the literature, it is known that the PEF inhibits (or accelerates) the enzymatic activity. However, most of the studies reported in the literature have been carried out on commercial enzymes or model foods. Nevertheless, this work was aimed to evaluate whether the electric field could decrease (or increase) the activity of the PPO enzyme, in both, the enzyme in real food (avocado paste) or on a commercial enzyme.

To generate the appropriate experimental conditions to treat the avocado pulp, it was necessary to develop previous experiments, on a commercial enzyme and an enzymatic extract of the avocado, in these experiments a sweep was carried out with the different frequencies to choose the best range of application of this variable. Also, an experiment was generated to find the best conditions to obtain the enzyme in terms of concentration and work volume to apply in the study of puree. The results of these two experiments are not described. However, they are implicit in the following experiment, and they constitute the experimental work of a Ph.D. thesis (Castorena García 2013).

2.1.2. Methodology

As mentioned, the virtue of the present study is the application of the electric field on the avocado puree, to inhibit the native polyphenol oxidase of the fruit. The puree, made of mixtures of 100 g of avocado pulp plus 30 mL of deionized water, homogenized with a blade blender, then divided into 60 g portions and placed in Petri dishes at -75 ºC in NUAIRE ultra-deep freezer until used.

A factorial experimental design with an arrangement (5x5), with treatment times (3, 6, 9, 12, and 15 minutes) and frequencies of (100, 260, 540, 730 and 950 Hz), plus the control sample with zero minutes of treatment and zero frequency. The constant conditions of the experiment were 8 g of puree placed in a Petri dish, placed between the two electrodes at 20 kV and 2 cm distance (10 kV • cm^{-1}), the square wave in bipolar mode, and five mA of electrical current.

Five grams of every treated sample was used to extract the PPO enzyme. To the treated puree 10 ml buffer Tris-HCl pH 7.1, 0.1 molars were added, the mixture was placed in 50 mL screw tubes, which were centrifuged in Eppendorf model 5804 R equipment at 11,000 rpm for 50 minutes and four ºC. The supernatant contains the active enzyme, and for its recovery, it was filtered on Whatman 40 medium pore paper and placed on ice until used.

The response to the EF effect was the measurement of enzyme activity on the substrate, for which 0.2 mL of the enzyme extract was taken and added to the substrate 3 mL of 5 millimolar catechol, (catechol (CAS 120-80- 9), dissolved in a base trizma (SIGMA-ALDRICH, Inc. St Louis Mo, USA), and adjusted with commercial Buffer of sodium phosphate and potassium phosphate pH 7.0 (No. 21231 of Hycel), and stopped with the addition of 0.6 ml of 35% hydrochloric acid (IDEQ, Puebla, Mexico).

The absorbance of the substrate-enzyme reaction was measured spectrophotometrically every 10 seconds for 1 minute at 420 nm in a UV-Vis spectrophotometer (Hewlett Packard model 8453 Germany), equipped with software for data acquisition, HP UV-visible ChemStation.

The activity of the polyphenol oxidase, was calculated by linear regression of the absorbance-time curve, the method was adjusted based on reported by González et al., (1999), Soliva-Fortuny et al., (2002), Arévalo et al., (2002), Gómez-López, (2002), Aquino-Bolaños and Mercado-Silva, (2004), Phunchaisri and Apichartsrangkoon, (2005), Dogan et al., (2005) and Koukounaras et al., (2008). The absorbance reading was used to calculate the enzymatic activity (S treated); this is compared with the activity of the untreated sample (S untreated) to evaluate the residual activity (RA), according to the following expression:

$$\% RA = \frac{S\ treated}{S\ untreated}\ 100$$

The results were analyzed through the analysis of variance (ANOVA) with a level of significance of 5% to evaluate the importance of each factor and its possible interactions, and then the Tukey test was used to know the most significant differences. Statistical analysis was performed with the Statistical Analysis System for Windows, SAS Ver 8 program (The SAS Institute Inc. Cary, NC., USA).

2.1.3. Results

The results of residual enzymatic activity of polyphenol oxidase of the different treated avocado puree extracts are shown in Table 2. Firstly, it is observed that there are frequencies, or experimental conditions, that far from inhibiting the activity of the enzyme seems to activate it, showing higher values of enzymatic activity revealed by the untreated enzyme. These effects have also been reported with the application of PEF (Giner et al., 2001). In the previous experiment on the commercial enzyme, it was found that it is better to work with frequencies lower than 1,000 Hz since, at very high frequencies, the effect is insensitive, and it is not detected.

Concerning the decrease in activity, Yeom et al., (2000) suggested that the effect of PEF produces conformational changes in the secondary structure of enzymes treated so that it could be responsible for modifications of the activity of the enzyme (Yeom et al., 2000).

Table 2. Results of the average residual activity

min	Hz	RA %*a	D std ±	min	Hz	RA %*b	D std ± .
3	25	88.5	13.44	3	540	91.5	3.54
6	25	75.5	12.02	6	540	75.0	4.24
9	25	71.5	4.95	9	540	68.5	6.36
12	25	61.5	7.78	12	540	64.5	19.09
15	25	56.0	5.66	15	540	69.5	17.68
3	100	98.5	2.12	3	730	94.0	0.00
6	100	90.0	0.00	6	730	79.0	2.83
9	100	80.0	12.73	9	730	66.0	2.83
12	100	68.0	19.80	12	730	63.5	0.71
15	100	58.5	21.92	15	730	56.0	4.24
3	260	110.0	7.78	3	950	92.0	7.07
6	260	70.0	9.90	6	950	80.0	11.31
9	260	81.5	13.44	9	950	75.0	2.83
12	260	78.5	13.44	12	950	67.5	2.12
15	260	77.8	17.68	15	950	58.5	10.61

* Average of 3 repetitions, *aElectric Field of 10 kV • cm^{-1}, *bElectric Field of 8kV • cm^{-1}, D std standard deviation, RA % Residual activity %.

The maximum inhibition of the enzyme polyphenol oxidase for the avocado paste was 44% (RA = 56), it was obtained at 25 and 730 Hz, both at 15 minutes of treatment; this RA value is within the range reported in the literature with different treatment systems and different operating conditions. In the preliminary experiment, not reported here, maximum inhibition of 88% was found for the commercial enzyme, or that is not in a food matrix, when treated with 10 kV/cm, 6 minutes at 260 Hertz; some others milder treatments activated the enzyme at values higher than 10%. From the results reported in Table 2, it is observed that activation of polyphenol oxidase (RA > 100%) was recorded in the treatment with 3 min and 260 Hz. The overall average effect of enzymatic inactivation recorded for this experiment was 24% (RA = 76). The Anova resulted significant for the two factors studied, the time was highly significant ($P < 0.001$) while the frequency was only significant ($P < 0.05$) the interaction was not significant indicating the independence of each of the factors. The correlation coefficient of the polynomial model was R = -0.87. In general, the effect of time is inversely proportional to RA, the higher the time of treatment, the higher the inhibition of the PPO enzyme. Figure 3 shows a summary of the percentage of RA in groups for each frequency and within this one bar for each treatment time.

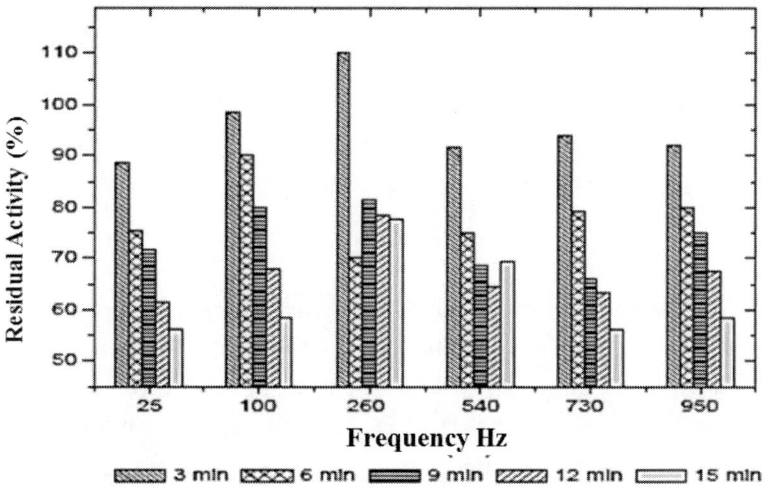

Figure 3. Residual enzymatic activity after different treatments with EF.

Figure 4. Petri-dishes with *samples* of EF-treated avocado (green) and untreated avocado (brown).

The horizontal scale is the frequency in Hz, and the height of the bar is related to the residual activity (RA). The height greater than 100% means activation of the enzyme, while the bars with lower value represent inhibition of the enzymatic activity. Treatment time has a more significant influence on the enzymatic inactivation than frequency; besides that, its effect is not proportional to the evaluated effect. Figure 4 shows a photograph with several Petri dishes with untreated and EF-treated samples.

Other emerging technologies, such as high hydrostatic pressure, have been used seeking the inactivation of polyphenol oxidase, for example, (López-Malo et al., 1999). They treated avocado puree at 345, 517, 689 MPa, per 10, 20 and 30 min and pH (4.3, 4.1, 3.9), they obtained 62% of residual activity at pH 3.9 and 20 min; at 345 MPa and 4.1 pH, 30 min, the residual activity obtained was 75%. Weemaes et al. (1998) worked with apple, avocado, grapes, and pears PFO; the treatment variables were 600, 700, 800 and 900 MPa and pH 6 to 7, the initial temperature was 25°C, rising when the working pressure increased. The inhibition of PFO was significant, and the effect was evaluated through a kinetic model of the first order. (Weemaes, et al., 1998).

2.1.4. Conclusion

According to the experimental results, it is possible to conclude that when the avocado puree is treated with EF, the activity of the polyphenol oxidase enzyme is reduced by an average of 24% and under certain conditions up to 44%; these percentages are similar to reductions in the enzymatic activity of other fruits or other non-thermal treatments. Only with thermal treatments is the enzyme 100% inhibited. With a 44% reduction in enzyme activity, it is possible to lengthen the shelf life of avocado puree; the most important of all this is that this reduction in the enzymatic activity that causes the darkening of the avocado has been done without raising the temperature of the purée. It is highly probable that, by increasing the treatment time, or the intensity, or the strength (the amperage), more reductions in this enzyme could be obtained.

3. Study of the Effect of the Electric Field on the Isomers of Avocado's Fatty Acids

3.1. Introduction

In the previous study, the effect of the electric field on the activity of the poly-phenol-oxidase enzyme was demonstrated, however, the possible effect of this treatment on other compounds present in the avocado pulp was not known, especially the compounds that have bridges of hydrogen or double bonds that are more unstable. These bonds could be affected by the strength of the electric field, so the fundamental idea of this project was to evaluate the possible formation of *trans* isomers in avocado oil after being subjected to the effect of the electric field. As it is known, in the last years, the harmful effect on the health of the *trans* isomers has been demonstrated, which are formed when the vegetable oil is subjected to high temperatures.

Likewise, several types of research have pointed out that the traditional methods to obtain avocado products diminish their sensory properties, in the same way, extraction methods to obtain oils generate trans-fatty acids, and that their consumption may have a detrimental effect on health since these have a similar effect to saturated fatty acids.

The central part of the present study is to expose the fatty acids and the pulp of the avocado to the effect of the high voltage electric field, for which two strategies were used: i) extract the oil, treat it with the electric field and analyze it, ii) treat the avocado pulp, extract the oil and analyze it. In both cases, the electric field was established considering the voltage, frequency, and time conditions with which the activity of the polyphenol oxidase enzyme was reduced. For the analysis of fatty acids, Gas Chromatography (GC) was used.

3.1.1. Methodology

Hass variety of avocado (*Persea Americana Mill*) from the State of Puebla, Mexico, was used. The fruit was washed with soap and water, and the husk and seed were removed, and the pulp was obtained, and their proximal analysis performed to characterize it, using the techniques of the AOAC (2009), each analysis was performed in triplicate.

Since it has been reported that the solvent extraction can also generate *trans*-isomers, avocado oil extraction was performed with solvent at room temperature by placing the sample inside a cartridge and subsequently immersed in the hexane inside a closed bottle for 24 hours. Another extraction was performed with solvent (Hexane), by the Soxhlet method 963.15 AOAC (2009) at 60°C for 4 hours, and oil extraction by centrifugation at 1157 g, for 10 minutes at 20, 25, 30, 35 and 40°C (Centrifuge 5804 R Eppendorf). (AOAC 2009).

The quantification of triacylglycerides was carried out by adding 20 mL of methanolic KOH to the oil. The solution refluxed at 60°C for 15 minutes. The methyl esters extracted with 10 mL of heptane, and 10 mL of distilled water was added; the two phases were separated. The residue was solubilized in heptane (IUPAC, 1990). 2 μL of the esters were taken and injected into a chromatographic column (RT-2560) of 100 μm length, 0.25 mm in diameter and 0.25 μm thick as support of fused silica biscyanopropyl of polysiloxane (Restek Corporation) coupled to a flame ionization detector (Hautfenne 1990).

The injector and the detector were maintained at 230 and 250°C, respectively. The carrier gas was nitrogen 1.2 mL/min with a division injection system (20:2). A mixture of Methyl Esters (Food Industry FAME Mix), with 37 components, was used to calibrate the Gas Chromatography system. They were identified by gas chromatography according to their retention time, and their comparison with a standard in an HP-5890 chromatograph (Hewlett-Packard Company, Wilmington, DE) equipped with a flame ionization detector.

The following strategy was developed to study the effect of the electric field on avocado oil:

1) The oil obtained by the three study methods was treated directly with the electric field.

2) The Avocado Pulp was exposed to the Electric Field; later on, the oil was extracted with the same three extraction methods proposed.

To corroborate the effect of the Electric Field in the avocado oil extracted by centrifugation a comparison took place; with an extra virgin olive oil (commercial brand), according to some researches, the olive oil contains a significant concentration of oleic acid (C18:1), similar to avocado oil, these obtained samples were treated with the Electric Field.

The conditions of the 2^3 factorial experimental design are shown in Table 3. The response variable was the formation of trans-fatty acids, the means of the treatments were compared by T-test for avocado oils extracted by the three proposed methods and an analysis of variance for avocado and olive oils ($\alpha = 0.05$).

Table 3. Experimental design of avocado oil treated to EF

Treatment	Voltage (kV/cm)	Frequency (Hz)	Time (s)
A	3	60	10
B	3	60	180
C	3	720	10
D	3	720	180
E	9	60	10
F	9	60	180
G	9	720	10
H	9	720	180
Control	-	-	-

3.1.2. Results

Table 4 shows the result of the bromatological analysis of the avocado pulp on a wet basis.

Separate investigations on avocado varieties have indicated that the amounts of various components are related to the growing area, growth conditions, collection time, and storage; these characteristics are unique for each variety. (C. Requejo-Tapia 1999, Gaydou, Lozano and Ratovohery 1987).

Table 4. Composition of the Hass avocado pulp used in the study

Component	%
Humidity	63.8 ± 3.1
Protein	1.4 ± 0.02
Lipids	25.3 ± 2.2
Fiber	0.51 ± 0.01
Ashes	1.5 ± 0.02
Carbohydrates	7.49 ± 1.2

Table 5. Chemical analysis of avocado oils, and Its comparison with other oils

Index	HOc	HO	CO	HAO	FAO	EOO	NMX
AI	1.29 ± 0.02	0.86 ± 0.02	0.75 ± 0.05	0.65	2.45	3.3	1.5
PI	3.79 ± 0.04	2.31 ± 0.11	2.46 ± 0.03	10.68	1.40	≤20	10
YI	84.65 ± 0.09	82.12 ± 0.06	86.67 ± 2.14	82.1	77.6	94	90

Table 5 shows the results of the chemical indices of avocado oils. Firstly, the extracted with hexane at 60°C (HOc); then extracted at room temperature (HO); afterward, extracted by centrifugation (CO); for comparison with Hass avocado oil (HAO), (Ortiz et al., 2003); Fuerte avocado oil (FAO), (Bora et al., 2001); extra-virgin olive oil (EOO), (CODEX, 1981); and maximum values established in accordance to the Mexican Standard, (SECOFI, 2008).

The results of Table 5 are expressed as AI: Acidity index expressed as a percentage of oleic acid. PI: Peroxide Index ($Meqkg^{-1}$). YI: Iodine index (analysis using the Wijs reagent).

From the previous results, the acid value in the oil extracted with hexane at 60°C has a higher value compared to that obtained by centrifugation and with solvent at room temperature. Bora et al., (2001), reported an acidity index of 2.45% in avocado oil extracted with hexane. Therefore, the results obtained in this work are lower, probably because it is another variety. However, if the results obtained for the acidity index are compared with the Mexican Standard, the three oils extracted by the different methods are within the permitted values, which is 1.5 expressed as a percentage of oleic acid.

Concerning the peroxide index, the oil extracted with hexane increasing the temperature presented a higher value than the oil obtained by the centrifugation method, and without heat, when comparing the results with that reported by the Mexican Standard, it could be seen that they are below the allowed value. Besides, these values were lower compared to those reported by Ortiz et al. (2003), Bora et al. (2001), and Virgin Olive Oil ≤ 20 (CODEX, 1981).

The results obtained for the Iodine Index are within the parameters reported by the Mexican Standard (SECOFI, 2008) this indicates that the fatty acid compositions in avocado and olive oils are very similar. From the previous results, it is observed that the best extraction method is by centrifugation because it maintained its characteristic green color, a solvent is not needed to extract it, and the fatty acids do not oxidize significantly, although the yield is remarkably low. Figure 5 shows the color of the avocado oils extracted by the three proposed methods.

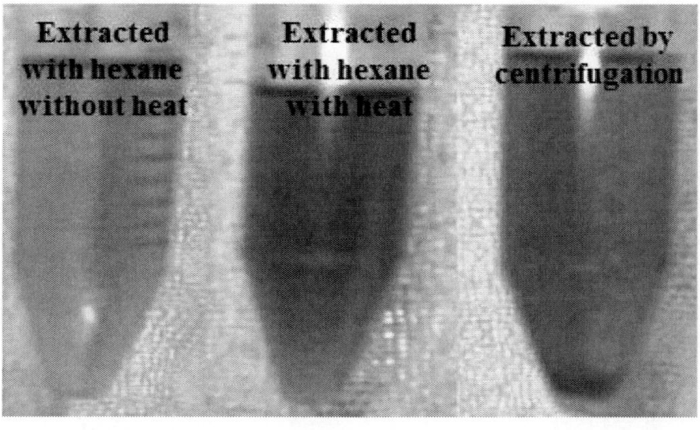

Figure 5. Visual comparison of avocado oils.

From the chromatograms, the peaks of each methyl ester included in the standard were identified, as well as their residence time at the operating conditions, each peak labeled with the initials of the corresponding fatty acid, for example, C4:0 to indicate acid butyric.

As in the chromatogram of the standard, the peaks of the oils studied were correlated with their retention times for proper identification. A total of 9 methyl esters were identified and quantified in the oil extracted without heat, 14 methyl esters in the oil extracted with heat and 12 methyl esters in the oil extracted by centrifugation.

The methyl ester that predominated was C18:1 with a percentage of 58.7 for the oil obtained by centrifugation, of 59.1, and 60.6% for oils extracted with hexane at 60 ° C at room temperature respectively.

Comparing these variations with the results reported by Ortiz et al., (2003), who analyzed the methyl esters of fatty acids from avocado oil of the Hass variety obtained by heating with microwaves, concluded that the C18:1 with 60.28% is in higher concentration. Figures 6, 7, 8, and 9 show the chromatograms of the standard and oil obtained with hexane, and at room temperature, the oil obtained with hexane at 60 ° C and that obtained through centrifugation, respectively.

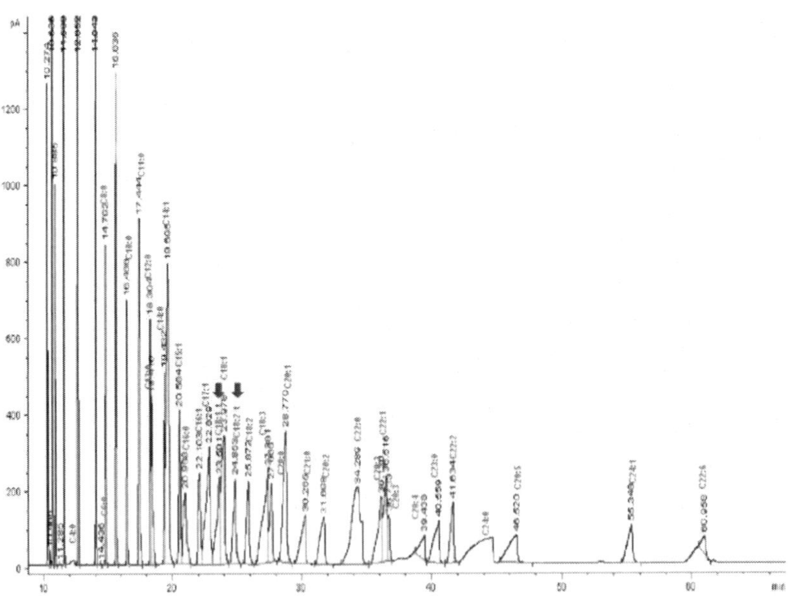

Figure 6. The chromatogram obtained from the Methyl Esters of the Standard.

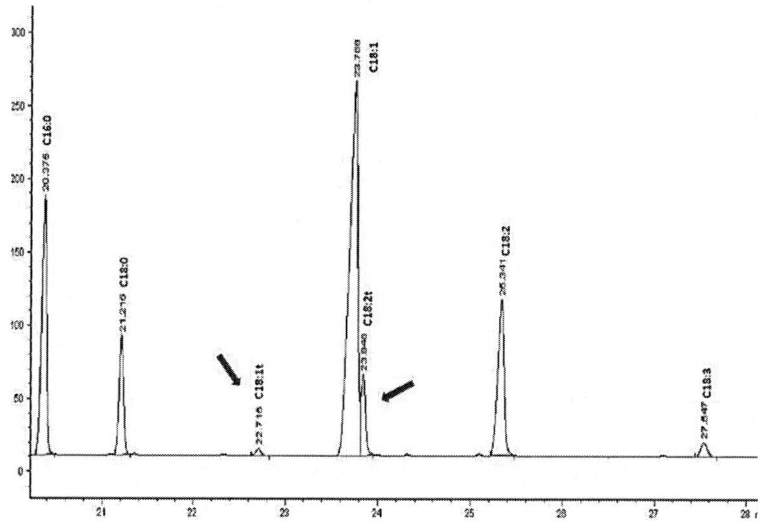

Figure 7. Chromatogram of oil extracted with hexane at room temperature.

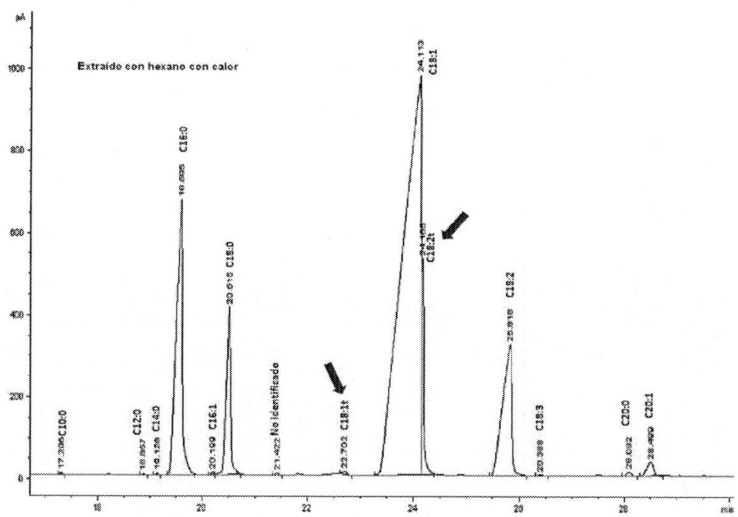

Figure 8. Chromatogram of oil extracted with hexane at 60°C.

The summary of the Methyl-Ester of Avocado-Oil, ME-AO, obtained by the three methodologies described before, as a percentage, is presented in Table 6 and compared with that reported by Ortiz et al., 2003. (M. Ortiz, et al. 2003).

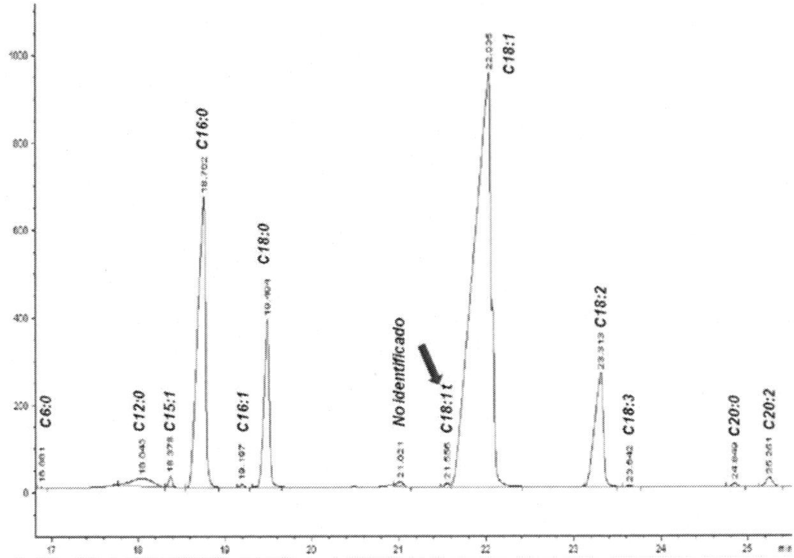

Figure 9. Chromatogram of oil extracted by centrifugation.

Table 6. Methyl esters of the fatty acids of avocado oils expressed in %

ME-AO	Extracted by centrifugation	Extracted with hexane TA.	Extracted with hexane 60°C	Microwave (Ortiz et al., 2003)
C10:0	0.02 ± 0.001	-	0.04 ± 0.002	-
C12:0	1.3 ± 0.05	-	0.02 ± 0.001	-
C14:0	0.31 ± 0.01	0.01 ± 0.001	0.04 ± 0.001	-
C16:0	17.06 ± 0.2	17.0 ± 1.1	13.7 ± 0.1	15.71
C16:1	7.1 ± 0.6	-	8.0 ± 0.2	7.26
C18:0	6.9 ± 0.4	5.1 ± 0.1	5.5 ± 0.1	0.72
C18:1t	0.1 ± 0.01	0.1 ± 0.01	0.1 ± 0.01	0.3
C18:1	58.7 ± 0.2	60.6 ± 2.1	59.1 ± 0.4	60.28
C18:2t	-	1.2 ± 0.1	3.1 ± 0.1	0.04
C18:2	7.3 ± 0.3	10.3 ± 1.2	9.2 ± 1.1	13.66
C18:3	0.06 ± 0.007	0.1 ± 0.01	0.04 ± 0.001	1.44
C20:0	0.1 ± 0.02	0.9 ± 0.02	0.17 ± 0.01	-
C20:1	-	-	0.17 ± 0.02	0.21
C20:2	0.48 ± 0.008	-	0.73 ± 0.01	0.11

Concerning trans-Fatty Acids, it is observed that there is a 0.1% C18:1t in the three oils, which indicates that it is probably found in the avocado oil natively.

With the hexane extraction methods, trans-Fatty Acids increased by 3.1 and 1.2% of C18:2t, respectively, observing the absence of these in the oil obtained by centrifugation, implying that the extraction method could influence the extraction of these fatty acids or in the formation of these. Wolff (1993) made a comparison of the trans isomers of α-linolenic acid (C18:3t) between the refined and cold-pressed oils of soybean and rape from different countries (Belgium, Great Britain, and Germany), finding that those obtained by Pressed samples showed traces of C18:3t, and for refined samples, they contained 1.0 to 3.3%, similar to that obtained in the present study.

In another study, Ortiz et al., (2003), extracted avocado oil with different methods and concluded that the method that generated the lowest amount of trans-Fatty acids was with the method that combines microwave and pressing with 0.33 of C18:1t and 0.04 g/100g of C18:2t.

In the tables presented below are the results regarding the composition of the ME-AO of the oils described in the experiment design (Table 3). The results presented correspond to the oils extracted before the EF treatment. Table 7 shows the results of the oils that were first extracted with hexane at room temperature and then treated with the electric field, EF. While Table 8 shows the results of the EM-AG, the oils obtained from avocado paste treated with EF and its subsequent extraction of oil with hexane at room temperature. In Table 7, it is observed that no treatment caused changes in C18:1t trans-Fatty acids, nor were there significant differences ($p > 0.05$) when comparing the results with the control oil. However, respect the concentration of the *trans* isomers, C18:2t, (methyl linoleaidate), treatments C (3 kV/cm, 720 Hz and 10 s), D (3 kV/cm, 720 Hz and 180 s), E (9 kV/cm, 60 Hz and 10 s) and F (9 kV/cm, 60 Hz and 180 s) they do were different ($p \leq 0.05$) to the control. Therefore, it is likely that the single extraction with hexane causes these differences in concentration. Each molecule in the oil phase has its defined configuration;

however, when it is dissolved in hexane, it surely has modifications in its structure that somehow modify the initial concentrations.

In Table 8, the avocado paste was first treated with the field and then extracted under the same extraction conditions as in Table 7. Respect the C18:1t, it can be observed that all the treatments, the results of this fatty acid, were minor or similar to the oil used as a control. Respect the other fatty acid, C18:2t, it is observed that in treatment B, the concentration rose slightly. However, there was a decrease ($p \leq 0.05$) of this fatty acid in treatments D to H. The common denominator of the treatments E, F, G, H is 9 kV/cm, treatment D is only 3 kV/cm, but longer time, it is very likely that the oil in the treated pulp has changed in its structure to give lower concentrations of this fatty acid, C18:2t.

Table 7. Methyl esters of the fatty acids of avocado oils obtained with hexane at room temperature and subsequent treatment with EF, in%

ME-AO	**A**	**B**	**C**	**D**	**Untreated**
C14:0	0.05 ± 0.004	0.06 ± 0.004	0.06 ± 0.001	0.05 ± 0.005	0.01
C16:0	16.1 ± 1.3	18.1 ± 1.4	17.7 ± 1.0	17.5 ± 1.4	11.0
C18:0	7.0 ± 0.06	7.0 ± 0.55	7.1 ± 0.4	7.1 ± 0.6	5.1
C18:1t	0.1 ± 0.03	0.1 ± 0.02	0.1 ± 0.09	0.1 ± 0.03	0.1
C18:1	42.1 ± 3.5	46.0 ± 3.5	43.3 ± 2.6	45.6 ± 3.5	40.6
C18:2t	0.1 ± 0.03	0.1 ± 0.02	0.4 ± 0.02	0.7 ± 0.03	0.2
C18:2	12.0 ± 1.1	12.6 ± 1.1	13.8 ± 0.8	14.3 ± 1.0	9.3
C18:3	0.1 ± 0.09	0.1 ± 0.01	0.1 ± 0.07	0.2 ± 0.01	.1
C20:0	1.0 ± 0.01	1.1 ± 0.09	1.2 ± 0.1	1.2 ± 0.09	0.9
ME-AO	**E**	**F**	**G**	**H**	**Untreated**
C14:0	0.05 ± 0.001	0.06 ± 0.001	0.06 ± 0.001	0.2 ± 0.04	0.01
C16:0	16.8 ± 1.3	18.5 ± 1.4	17.6 ± 0.01	13.4 ± 1.3	11.0
C18:0	6.6 ± 0.5	7.1 ± 0.54	7.1 ± 1.3	5.2 ± 0.5	5.1
C18:1t	0.1 ± 0.01	0.1 ± 0.03	0.1 ± 0.03	0.1 ± 0.02	0.1
C18:1	43.3 ± 3.4	46.1 ± 3.5	45.1 ± 3.2	43.6 ± 3.5	40.6
C18:2t	0.7 ± 0.03	0.7 ± 0.02	0.1 ± 0.04	0.2 ± 0.03	0.2
C18:2	12.9 ± 0.04	13.4 ± 0.97	14.1 ± 1.2	10.3 ± 1.09	9.3
C18:3	0.1 ± 0.01	0.2 ± 0.01	0.1 ± 0.01	0.2 ± 0.01	0.1
C20:0	1.1 ± 0.09	1.0 ± 0.08	1.2 ± 0.11	1.0 ± 0.09	0.9

Table 8. Methyl esters of avocado oil obtained from paste treated with EF and its subsequent extraction with hexane at room temperature (%)

ME-AO	A	B	C	D	Untreated
C16:0	19.3 ± 1.4	19.8 ± 1.3	19.5 ± 1.4	19.2 ± 1.5	17.2
C18:0	8.6 ± 0.6	10.2 ± 0.7	7.8 ± 0.6	9.3 ± 0.6	7.1
C18:1t	0.6 ± 0.04	0.4 ± 0.05	0.6 ± 0.03	0.6 ± 0.05	0.6
C18:1	43.8 ± 3.4	40.1 ± 3.4	45.7 ± 3.5	43.2 ± 3.5	40.4
C18:2t	0.5 ± 0.02	0.6 ± 0.04	0.5 ± 0.04	0.4 ± 0.04	0.5
C18:2	11.0 ± 0.8	12.1 ± 1.0	10.3 ± 0.02	11.6 ± 0.8	10.2
C18:3	0.9 ± 0.07	0.9 ± 0.01	0.8 ± 0.08	1.0 ± 0.06	0.9
ME-AO	**E**	**F**	**G**	**H**	**Untreated**
C16:0	19.1 ± 1.4	19.3 ± 1.5	19.6 ± 1.7	19.4 ± 1.5	17.2
C18:0	8.3 ± 0.06	8.2 ± 0.08	8.1 ± 0.63	8.3 ± 0.07	7.1
C18:1t	0.6 ± 0.04	0.5 ± 0.03	0.6 ± 0.04	0.5 ± 0.05	0.6
C18:1	44.8 ± 3.4	44.8 ± 3.1	45.0 ± 3.5	45.1 ± 3.4	40.4
C18:2t	0.3 ± 0.04	0.3 ± 0.05	0.2 ± 0.04	0.3 ± 0.04	0.5
C18:2	10.8 ± 0.8	10.5 ± 0.9	10.8 ± 0.3	10.7 ± 0.8	10.2
C18:3	0.9 ± 0.02	0.8 ± 0.07	0.8 ± 0.07	0.9 ± 0.07	0.9

In conclusion, despite the EF treatment, the oil extracted by this method is within the maximum limit allowed by international standards, as established by the Danish Ministry of Health, with a maximum of 1 g/100 g of lipid. On the other hand, the FDA allows contents less than 0.5 g/100 g, in addition to which the quantified values were lower with some of the treatments when compared with an untreated sample.

In the sequence of experiments carried out, the following analysis presents the results of the oil extracted with hexane, but at a temperature of 60°C and subsequently treated with the EF, (Table 9).

The effect of the temperature in the extraction, before the EF treatment is evident, in the two trans-Fatty acids (FA) being analyzed, all treatments show higher amounts of these trans-FA. Additionally, it is observed that the extraction with hexane at 60°C, increases the content of C18:1t and C18:2t, and if treatment is also given with EF as a conservation method, it increases even more, since all the results present significant differences with the control sample, ($p \leq 0.05$).

Table 9. Methyl esters of avocado oil extracted with hexane at 60°C with subsequent treatment with EF expressed in%

ME-AO	A	B	C	D	Untreated
C16:0	19.07 ± 1.5	19.14 ± 1.5	19.0 ± 1.4	19.0 ± 1.4	19.0
C18:0	7.9 ± 0.6	7.3 ± 0.6	8.1 ± 0.6	8.3 ± 0.6	6.2
C18:1t	0.54 ± 0.04	0.59 ± 0.03	0.5 ± 0.04	0.3 ± 0.04	0.1
C18:1	43.7 ± 3.3	43.7 ± 3.4	43.3 ± 3.4	43.6 ± 3.3	40.6
C18:2t	4.6 ± 0.3	3.8 ± 0.3	4.7 ± 0.3	4.0 ± 0.4	3.5
C18:2	12.9 ± 1.0	12.2 ± 0.2	13.0 ± 1.0	13.1 ± 1.0	12.0
C18:3	1.2 ± 0.09	1.2 ± 0.1	1.2 ± 0.1	1.2 ± 0.09	0.6
C20:0	-	-	-	-	-
ME-AO	**E**	**F**	**G**	**H**	**Untreated**
C16:0	19.4 ± 1.5	19.4 ± 1.5	19.1 ± 1.4	18.5 ± 1.5	19.0
C18:0	7.4 ± 0.6	7.2 ± 0.6	8.0 ± 0.5	8.1 ± 0.1	6.2
C18:1t	0.4 ± 0.03	0.4 ± 0.04	0.3 ± 0.03	0.5 ± 0.06	0.1
C18:1	45.7 ± 3.5	44.6 ± 3.6	46.2 ± 3.6	45.4 ± 3.4	40.6
C18:2t	3.4 ± 0.2	3.5 ± 0.3	4.4 ± 0.3	4.6 ± 0.4	3.5
C18:2	12.2 ± 0.9	11.9 ± 0.9	13.0 ± 0.9	13.5 ± 1.0	12.0
C18:3	1.2 ± 0.09	1.2 ± 0.08	1.2 ± 0.06	1.3 ± 0.09	0.6
C20:0	1.1 ± 0.08	1.04 ± 0.01	-	-	-

For the case of pulp treated first with the EF and then extracted the oil with hexane at 60°C, Table 10 shows the results.

For the C18:2t fatty acid, it can be seen that with two exceptions, treatments E and H, the rest of the treatments had significant increases in the concentration of that trans-Fatty acid about the control sample. For the C18:1t fatty acid, all treatments had higher amounts of this FA than the control. With these results, it could be said that heat stimulates the formation of trans-Fatty acids and that, in addition to applying the EF, they increase more; hence, this procedure is not recommended to conserve this oil. However, it is essential to note that the values for trans-Fatty acids are lower here than in the Table, indicating that somehow the damage is less when the pasta is first treated with the EF.

Table 10. Methyl esters of avocado oil obtained from avocado paste treated with EF and then extracted with hexane at 60°C (%)

ME-AO	A	B	C	D	Untreated
C14:0	0.05 ± 0.004	0.02 ± 0.01	-	-	-
C16:0	20.1 ± 1.3	19.8 ± 1.4	20.0 ± 1.5	18.8 ± 1.5	19.1
C18:0	9.39 ± 0.7	10.2 ± 0.7	9.3 ± 0.8	9.2 ± 0.7	6.3
C18:1t	0.27 ± 0.03	0.47 ± 0.04	0.3 ± 0.01	0.18 ± 0.1	0.1
C18:1	33.6 ± 2.6	40.1 ± 3.3	41.4 ± 3.3	39.6 ± 3.2	40.2
C18:2t	4.3 ± 0.3	6.2 ± 0.2	4.1 ± 0.2	3.3 ± 0.3	3.1
C18:2	12.19 ± 0.8	12.1 ± 0.9	10.6 ± 0.7	12.2 ± 0.2	11.0
C18:3	0.13 ± 0.06	0.972 ± 0.07	0.13 ± 0.04	1.2 ± 0.8	0.1
ME-AO	**E**	**F**	**G**	**H**	**Untreated**
C14:0	-	0.2 ± 0.001	0.01 ± 0.02	0.05 ± 0.004	-
C16:0	20.9 ± 1.6	19.9 ± 1.5	17.2 ± 1.5	19.5 ± 1.6	19.1
C18:1t	0.8 ± 0.01	0.53 ± 0.03	0.36 ± 0.1	0.16 ± 0.02	0.1
C18:1	45.8 ± 1.4	43.1 ± 3.1	42.3 ± 3.0	43.0 ± 3.2	40.2
C18:2t	2.0 ± 0.1	6.1 ± 0.5	4.7 ± 0.4	2.2 ± 0.3	3.1
C18:2	16.1 ± 0.2	11.9 ± 0.9	10.9 ± 0.8	12.2 ± 0.9	11.0
C18:3	1.0 ± 0.1	1.0 ± 0.07	0.8 ± 0.01	0.13 ± 0.01	0.1

Table 11. Methyl esters of avocado oil extracted by centrifugation and subsequent treatment with EF expressed in %

ME-AO	A	B	C	D	Untreated
C14:0	0.02 ± 0.007	0.02 ± 0.005	0.02 ± 0.002	0.02 ± 0.009	0.13
C16:0	16.0 ± 0.3	16.3 ± 0.18	17.0 ± 0.3	17.0 ± 0.11	17.0
C16:1	7.1 ± 0.2	7.07 ± 0.9	7.01 ± 0.8	7.04 ± 0.1	7.01
ME-AO	A	B	C	D	Untreated
C18:0	6.1 ± 0.5	6.7 ± 0.3	6.4 ± 0.1	7.1 ± 0.2	6.9
C18:1t	0.02 ± 0.003	0.01 ± 0.002	0.07 ± 0.003	0.06 ± 0.008	0.1
C18:1	57.0 ± 0.1	57.2 ± 0.1	58.1 ± 0.44	57.0 ± 0.1	58.7
C18:2t	0.1 ± 0.006	0.05 ± 0.001	0.1 ± 0.002	-	-
C18:2	8.4 ± 0.1	9.0 ± 0.1	9.0 ± 0.16	11.0 ± 0.11	7.3
C18:3	0.9 ± 0.001	0.3 ± 0.008	0.9 ± 0.005	0.1 ± 0.007	0.06
C20:0	1.0 ± 0.06	0.1 ± 0.007	1.1 ± 0.07	0.4 ± 0.002	0.16
ME-AO	**E**	**F**	**G**	**H**	**Untreated**
C14:0	0.14 ± 0.01	0.01 ± 0.008	0.07 ± 0.004	0.02 ± 0.003	0.13
C16:0	17.3 ± 0.13	16.6 ± 0.16	17.0 ± 3.09	17.0 ± 0.6	17.0
C16:1	7.06 ± 0.1	7.02 ± 0.3	7.02 ± 0.1	7.01 ± 0.1	7.01

Table 11. (Continued)

ME-AO	E	F	G	H	Untreated
C18:0	6.0 ± 0.11	7.0 ± 0.12	7.0 ± 0.27	6.3 ± 1.03	6.9
C18:1t	0.01 ± 0.002	0.01 ± 0.002	0.08 ± 0.003	0.07 ± 0.003	0.1
C18:1	59.0 ± 0.5	55.2 ± 1.8	58.4 ± 1.4	56.0 ± 2.5	58.7
C18:2t	0.03 ± 0.004	0.07 ± 0.001	-	0.09 ± 0.001	-
C18:2	8.4 ± 0.3	9.0 ± 0.8	10.2 ± 3.1	8.3 ± 0.5	7.3
C18:3	0.11 ± 0.002	0.10 ± 0.001	0.01 ± 0.001	0.9 ± 0.005	0.06
C20:0	0.13 ± 0.001	1.0 ± 0.07	0.3 ± 0.002	0.95 ± 0.007	0.16
C20:1	1.30 ± 0.08	0.7 ± 0.001	0.9 ± 0.001	0.9 ± 0.001	0.05

Table 12. Methyl esters of avocado oil obtained from pulp treated with EF and then extracted by centrifugation expressed in %

ME-AO	A	B	C	D	Untreated
C14:0	0.02 ± 0.007	0.02 ± 0.005	0.02 ± 0.002	0.02 ± 0.009	0.13
C16:0	16.0 ± 0.3	16.3 ± 0.18	17.0 ± 0.3	17.0 ± 0.11	17.0
C16:1	7.1 ± 0.2	7.07 ± 0.9	7.01 ± 0.8	7.04 ± 0.1	7.01
C18:0	6.1 ± 0.5	6.7 ± 0.3	6.4 ± 0.1	7.1 ± 0.2	6.9
C18:1t	0.02 ± 0.003	0.01 ± 0.002	0.07 ± 0.003	0.06 ± 0.008	0.1
C18:1	57.0 ± 0.1	57.2 ± 0.1	58.1 ± 0.44	57.0 ± 0.1	58.7
C18:2t	0.1 ± 0.006	0.05 ± 0.001	0.1 ± 0.002	-	-
C18:2	8.4 ± 0.1	9.0 ± 0.1	9.0 ± 0.16	11.0 ± 0.11	7.3
C18:3	0.9 ± 0.001	0.3 ± 0.008	0.9 ± 0.005	0.1 ± 0.007	0.06
C20:0	1.0 ± 0.06	0.1 ± 0.007	1.1 ± 0.07	0.4 ± 0.002	0.16
ME-AO	E	F	G	H	Untreated
C14:0	0.14 ± 0.01	0.01 ± 0.008	0.07 ± 0.004	0.02 ± 0.003	0.13
C16:0	17.3 ± 0.13	16.6 ± 0.16	17.0 ± 3.09	17.0 ± 0.6	17.0
C16:1	7.06 ± 0.1	7.02 ± 0.3	7.02 ± 0.1	7.01 ± 0.1	7.01
C18:0	6.0 ± 0.11	7.0 ± 0.12	7.0 ± 0.27	6.3 ± 1.03	6.9
C18:1t	0.01 ± 0.002	0.01 ± 0.002	0.08 ± 0.003	0.07 ± 0.003	0.1
C18:1	59.0 ± 0.5	55.2 ± 1.8	58.4 ± 1.4	56.0 ± 2.5	58.7
C18:2t	0.03 ± 0.004	0.07 ± 0.001	-	0.09 ± 0.001	-
C18:2	8.4 ± 0.3	9.0 ± 0.8	10.2 ± 3.1	8.3 ± 0.5	7.3
C18:3	0.11 ± 0.002	0.10 ± 0.001	0.01 ± 0.001	0.9 ± 0.005	0.06
C20:0	0.13 ± 0.001	1.0 ± 0.07	0.3 ± 0.002	0.95 ± 0.007	0.16

As for the oils obtained by centrifugation, the results are shown in Tables 11 and 12, in the first case the oil obtained by centrifugation and then treated with the EF, in the second case the oil obtained from the avocado paste treated with the EF and then extracted by centrifugation to separate the two phases.

In the previous table, the C18:1t trans isomers decrease their concentration; in all cases concerning the untreated sample, the C18:2 isomers, only treatments A and C show a value slightly different than the control.

In this case, the oil obtained by centrifugation of the treated paste is the smoothest method, since the concentration of the two trans-Fatty acids that were analyzed showed lower concentrations than the control (o untreated sample).

3.1.3. Conclusion

It is useful to remember that the application of the electric field on the avocado paste was sought partially inactivate the enzyme responsible for the enzymatic obscuration of the avocado. Under the experimental conditions where the enzyme had its inhibition, now we have shown that there are no negative consequences on other components of the avocado, such as the generation of trans isomers, which was minimal, and in some other cases even lowered. Therefore, the use of this technology is recommended to extend the shelf life of avocado paste. Even though, considering that the concentration of these compounds also depends on the extraction conditions of the oil; here it has been shown that the application of heat is a factor that increases the generation of the compounds, which may be harmful to the health of consumers, who seeks to reduce their consumption.

From the results of this project and as a conclusion the best method of obtaining avocado oil is by centrifugation because the avocado oil maintains its characteristic green color, solvents are not required to extract it, and its fatty acids do not present noticeable oxidation, according to the chemical analyses carried out. Besides, gas chromatography evaluated the presence of trans isomers in avocado oil and found that they are at naturally low concentrations, but these concentrations are within the maximum limit allowed by international standards.

3.2. Study of the Effect of the Electric Field on Vitamins C, E and A of Avocado

3.2.1. Introduction

When treating avocado paste with the electric field (EF), a substantial percentage is achieved in the inactivation of the polyphenol oxidase (Castorena, 2007). Moreover, the application of the field does not generate more trans-Fatty acids than those that the pulp naturally contains. In other studies, we have shown that the electric field produces a decrease in microbial contamination of food (Robles 2008). The goodness of the EF technique lies in its application for food preservation does not imply an increase in the temperature of the product during the treatment, which is beneficial for the product, since in general, it retains its nutritional and sensory characteristics. However, because this conservation technique is recent, it is unknown whether it has undesirable effects on other nutritional components of food.

This part of the work describes the results of the application of the electric field on the avocado paste to evaluate its effect on vitamins A, C, and E present in the fruit. In general, vitamins are sensitive to factors such as oxygen, light, temperatures, metal ions, among others, so it is possible that the EF treatment applied to the avocado paste may cause a degradation of the vitamins present.

3.2.2. Methodology

Because the vitamins in most of the natural products are in low concentrations, in this work, we established the strategy of applying the electric field and evaluating its effect on the standards of vitamins alone and then added into an avocado paste, it is, as the internal standard. The analysis of the vitamins was carried out, in general, utilizing high-performance liquid chromatography (HPLC) using the equipment: HP

1100 equipped with in-line degasser, quaternary pump, automatic injector, and diode array detectors (DAD) and Fluorescence (FLD). In the case of beta-carotene, it was also evaluated through a Genesys 2 UV/visible spectrophotometer (Madison, Wisconsin, USA). In the case of vitamin E, the infrared spectrometer with a Fourier transform was explored, (FTIR) using the equipment brand Bruker Vertex 70, (Riga LV-1658, Latvia); these analyses were performed by the method of Attenuated Total Reflectance (ATR) to assess whether in the future it is possible to facilitate the analysis of this vitamin. An Eppendorf brand centrifuge, model 5804 R, was also used (Madrid, Spain).

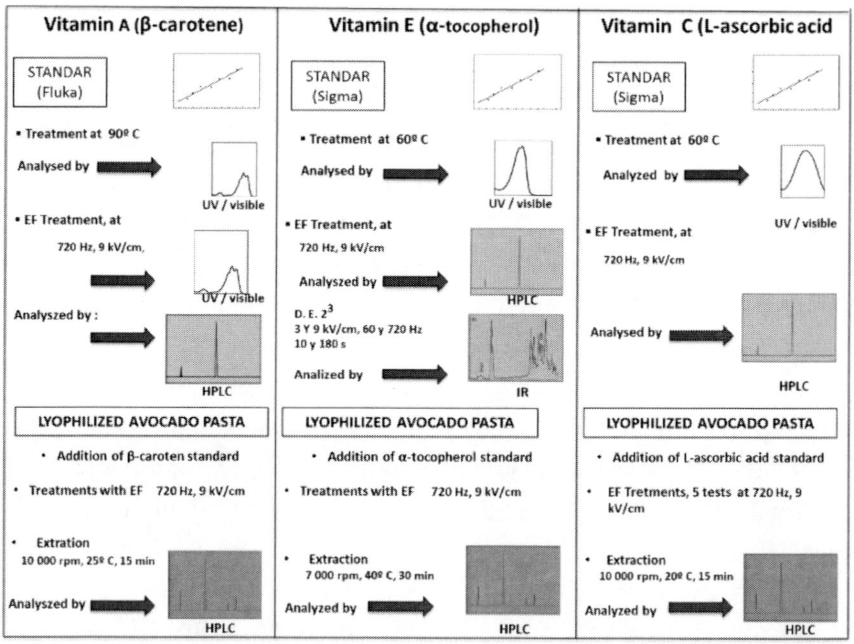

Figure 10. Experimental development of standards and avocado paste treated with EF.

The standards used were β-carotene from Fluka, L-ascorbic acid, and α-tocopherol from Sigma. Besides, reactants such as hexane, and Golden Bell isopropyl alcohol, acetonitrile, methanol from EMD Chemicals, ethyl acetate and water (HPLC grade); spectroscopic grade sulfuric acid from Sigma. Freeze-dried Hass avocado paste was obtained using a Virtis equipment (Warminster, PA, USA), to have the same batch in all the experimentation.

For the quantification of vitamins, calibration curves were constructed using different concentrations of the vitamin and the area under the curve of the peak produced in the HPLC equipment. The concentrations were chosen based on the amount of vitamin present in 100 g of avocado. The procedure consisted of preparing a stock solution of vitamin, from which the necessary dilutions were made for the curve.

For the treatments with the electric field, the same equipment that was designed and built in the CIBA-IPN was used, and that has already been described in the first part of this work.

Figure 10 summarizes the innovative strategy of the present work, in each of the three columns the methodology for each of the vitamins is described, which are the ones that are mostly found in the avocado according to the literature.

3.2.2.1. Application of the EF Electric Field to Standards

The EF treatments applied were at 9 kV/cm of voltage, 720 Hz of frequency, ant application times were between 0.5, up to 30 minutes, all the experiments were performed in duplicate, under a completely randomized design. All samples filtered before being injected into the chromatograph (HPLC), through a 0.45 μm pore membrane. The control samples were placed in the treatment chamber of the EF equipment for the same times defined in the experimental design, but without applying voltage.

3.2.2.2. Analysis in UV/Visible Spectroscopy to Quantify EF Effects on Standard-β-Carotene

The samples treated with EF analyzed in the spectrophotometer by scanning the absorption spectra, from 200 to 1100 nm, the spectra of the different treatments compared with each other, as well as against each control sample.

3.2.2.3. Analysis of β-Carotene in Avocado Paste Treated with CE in RP-HPLC

40 g of lyophilized avocado, 208 ml of water, and 32 ml of a 120 μg/ml solution of β-carotene standard in ethyl acetate were prepared. Samples of 17.5 g were taken, and the EF applied. The treated samples were centrifuged at 10,000 rpm at 25°C for 15 min, from the upper phase 0.1 g was taken and diluted to 1 ml with the mobile phase, from this dilution 20 μl (previously filtered) was injected. The samples were eluted with 60% acetonitrile, 30% ethyl acetate, and 10% methanol at a flow rate of 1 ml/min. The detection was made with the DAD at 450 nm.

3.2.2.4. Analysis of L-Ascorbic Acid Treated With EF in RP-HPLC

Samples of 1 ml of 150 μg/ml standard in 4% metaphosphoric acid were treated. Ten μl aliquots were injected into the HPLC using a C18 column of 250 x 4.6 mm, five μm Restek; the elution was with 0.05% H_2SO_4 with a flow of 0.8 ml/min. Detection was performed with DAD at 245 nm.

3.2.2.5. Analysis of L-ascorbic Acid in Avocado Paste Treated with EF in HPLC

A 175 g sample was prepared, with 14.3% lyophilized pulp, 68.6% water, and 0.043% of ascorbic acid. 15 g per sample was weighed. The ascorbic acid extracted by centrifuging the samples at 20°C at 10,000 rpm for 15 minutes, ten μl of the aqueous phase taken to be analyzed on HPLC.

3.2.2.6. Treatments for Avocado Paste with Internal Standards

The samples were prepared by mixing 2.5 g of lyophilized avocado with 15 ml of distilled water and 2 mg of α-tocopherol standard.

Twelve samples were prepared with 2.5 g of lyophilized pulp, 12 ml of water, and the internal standard of L-ascorbic acid, in a concentration of 0.43% diluted in 4% metaphosphoric acid.

The treatments were carried out with the same conditions performed for the treatments of the standards. They were centrifuged at 7,000 rpm at

40° C for 30 minutes, from the oily part, 20 µl were taken, and digested in 1 ml of hexane, a ten µl aliquot was used for chromatography. With an isocratic elution with 99% Hexane, 1% Isopropyl Alcohol, at a flow of 0.8 ml/min, the detection was made with the Ex Fluorescence detector: 290 nm, and with different wavelengths.

Analysis of variance (ANOVA) was performed to measure the significance of the treatments. The calibration curves were obtained from the linear regression analysis from the peak areas of the chromatograms against the concentrations of each of the standards (Excel 2003 from Microsoft). Likewise, for the adjustments to a model, it was calculated using linear regression of the transformed data.

3.2.3. Results

3.2.3.1. β-Carotene

Previous studies of the absorbance of beta-carotene showed that this compound presents a peak of maximum absorption at 450 nm, as long as this vitamin is being degraded (by heat, by the electric field and by oxidation), that peak decreases its intensity and another peak appears around the 340 nm. These values and the corresponding spectra are similar to those reported by Rodríguez-Amaya et al. (Rodríguez-Amaya 2004). Figure 11 shows the effect of the EF on the absorbance value of these two wavelengths as a function of the treatment time with the electric field.

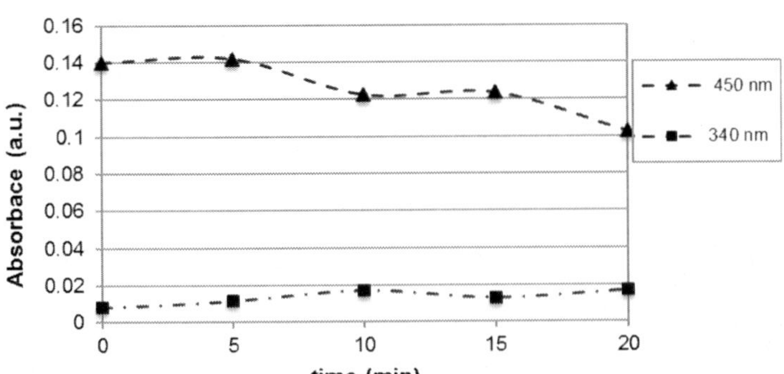

Figure 11. The absorbance of beta-carotene after treatment with the EF at two wavelengths.

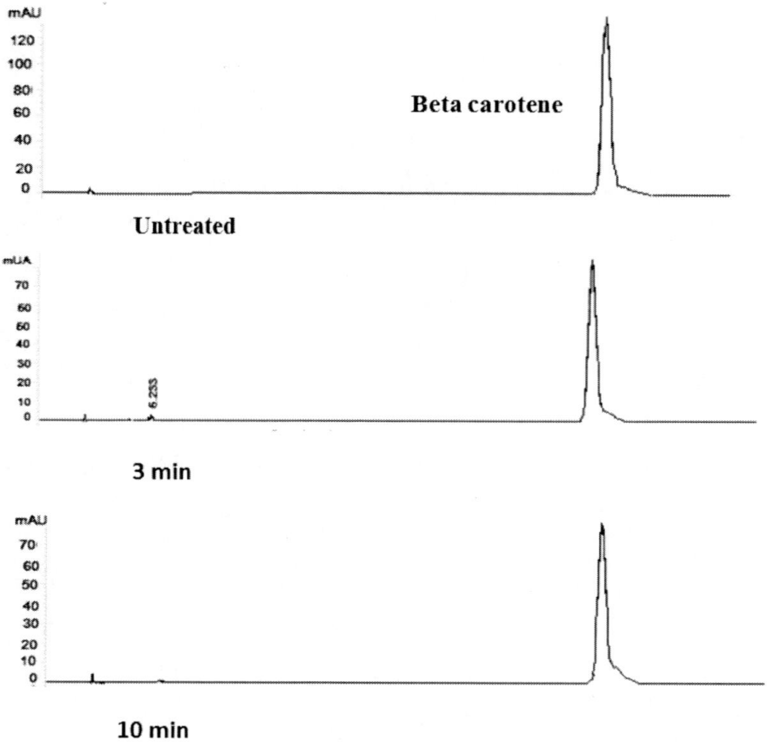

Figure 12. Chromatograms of β-carotene standard treated with EF (450 nm).

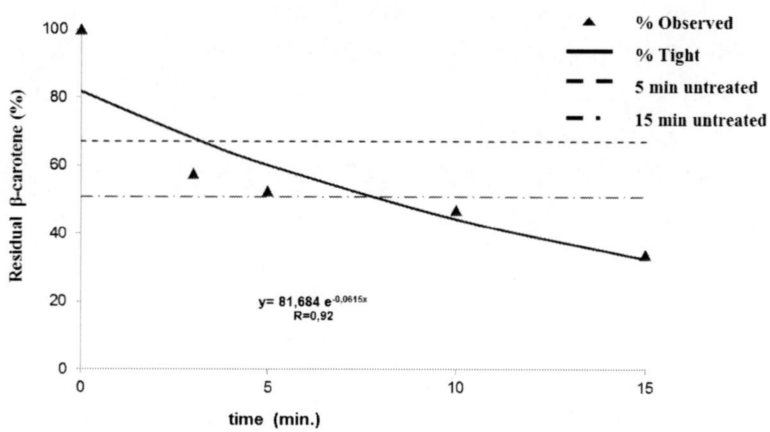

Figure 13. Percentage of residual β-carotene after treatment with EF.

Concerning the HPLC analysis, the data of the β-carotene type curve were adjusted to a linear model, R = 0.9968, with the adjustment equation, the β-carotene was quantified in the samples before and after the treatments with EF. Figure 12 shows a decrease in the absorbance of β-carotene, as well as the detection of a new peak at minute 6.2, in samples treated with EF for 15 minutes. In the literature it is mentioned that isomers such as 13, 15-di-cis or 9-cis-β-carotene present retention times slightly higher than β-carotene, or the peak of the cis compound are bound to the trans peak, (Rodríguez -Amaya 2004), (Rodriguez-Amaya and Azebedo-Meleiro 2004), (Davies, Austin and Partridge 1991), (Davey, Keulemans and Swennen 2006), (Qiu, Chen and Li 2009), (Qiu, Chen and Li 2009). In Figure 12, the chromatograms of the beta-carotene are shown, firstly, the untreated standard of the vitamin. Secondly, the standard treated at two different times.

These results coincide with the results of UV/visible spectroscopy, where the generation of cis isomers of β-carotene was also detected as an EF effect. It was observed that the degradation of β-carotene, β-cryptoxanthin and α-carotene increased with the intensity of the electric field. Quantification was performed before and after the treatments to assess the effect of the electric field on the standards of vitamins treated. The amount of vitamin detected after treatments are called "residual vitamin." Figure 13 shows the residual vitamin results after the treatments.

It is important to note that the controls of 5 and 15 minutes, who had no treatment, show a decrease in this vitamin due to the effect of the environment alone. A loss of about 30% of the vitamin occurred at 5 minutes, and 50% at 15 minutes, indicating that β-carotene had low stability in the environmental conditions present during the tests. These results also tell us that this vitamin is labile to the environment alone. On the other hand, in samples treated with EF, there are significant losses, although there is a reduction from 3 minutes of treatment (57%); hence, 44% is due to environmental effects, and only 13% is due to the effect of the EF. The data were adjusted to an exponential model with R = 0.92, where the degradation rate of β-carotene is 0.061 $minutes^{-1}$. The results of the ANOVA showed that there is a significant difference ($p < 0.05$)

between the control samples and the treated samples, which indicates that the EF on the standard does produce a loss of this vitamin, although the control samples also lost vitamin.

However, in the avocado paste, added with the vitamin, when treated with the EF, no new peaks were detected in the chromatographic results, in all the treatments of this test, only slight variations of the peak area of β-carotene were observed for the samples without treating.

The behavior of the vitamin under EF treatment was adjusted to an exponential model, where the degradation speed of β-carotene is now 0.01 minutes-1; that is to say approximately one-sixth of the degradation speed of the standard, with which it is observed that in all the treatments the decrease is much lower in comparison to that registered in the standard.

Besides, no significant differences were found ($p > 0.05$) between avocado paste treatments. Some studies have reported on the effect of PEF in the retention of carotenoids of juices, and in such cases, the decreases are zero, or the treatment improves the extraction of these compounds (Torregrosa, and others 2005). These examples show that the compound that is analyzed is in a food matrix, suggesting that, given that the avocado matrix is very complex, it protects β-carotene in some way, preventing its degradation.

3.2.3.2. α-tocopherol

For the case of this vitamin (E), the concentration of the same has a linear relationship with the area under the curve of the respective chromatogram; the linear regression yields a correlation coefficient R = 0.976. Figure 14 shows the effect of the electric field on the concentration of the internal standard alone. The figure is shown in terms of the percentage of residual alpha-tocopherol as a function of the treatment time. The results are presented for the two detectors used in the HPLC.

The previous figure shows how the concentration of alpha-tocopherol decreases and increases, even more than the control sample, only in one case the concentration decreased more significantly than 20%. The previous tells us that the environment, together with the electric field, are those that exert that variation concerning the control sample that had no treatment and was not exposed to the environment at the same time. When performing the statistical analysis of the samples, it was observed that there is no significant difference between the treatments ($p > 0.05$). This suggests that the electric field does not have a deteriorating effect on the alpha-tocopherol standard.

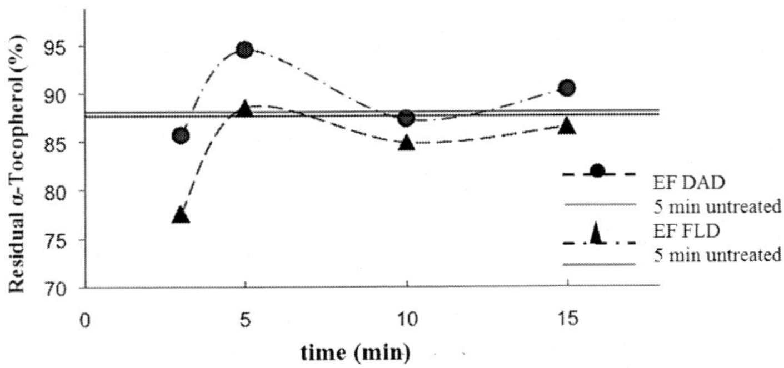

Figure 14. Percentage of residual α-tocopherol from samples treated with EF, with DAD (298 nm) and FLD (Ex = 290, Em = 330 nm).

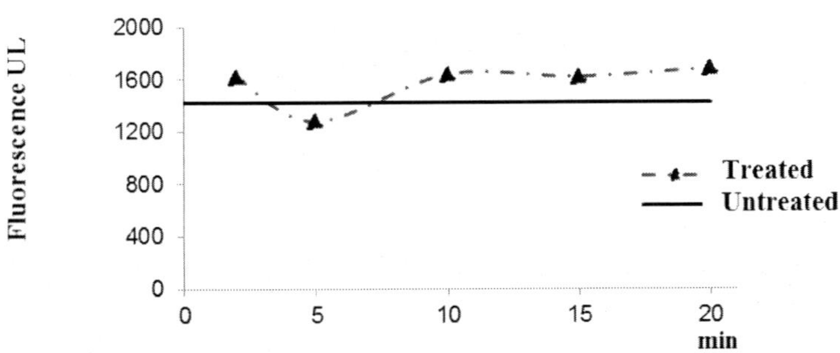

Figure 15. Dependence of the fluorescence of residual α-tocopherol in avocado pastes treated with EF (FLD, Ex = 290, Em = 330 nm).

Figure 15 shows the HPLC results, using the fluorescence detector, of the residual vitamin from the EF treated avocado paste added with the standard of α-tocopherol; the EF treatment was under the experiment design described above.

Figure 15 shows us that the damage caused by the EF, with times higher than 5 minutes, is less than the effect of the environment alone. It is also likely that EF could improve the extraction of tocopherol, as has been reported for the extraction of other vitamins (Odriozola-Serrano et al., 2009). The analysis of variance shows that there is no significant difference between the treatments ($p > 0.05$).

2.2.3.3. L-ascorbic Acid Treated with EF and HPLC Analysis

The adjustment of the area under the curve of the corresponding peak for this vitamin with the concentration used, through linear regression, produced a correlation coefficient R = 0.9994. The magnitude of the damage exerted by the electric field on the standard only of this vitamin is shown in Figure 16, in which the percentages of residual ascorbic acid of the treated and untreated samples are indicated.

In Figure 16, it is observed that the 10-minute control and the untreated sample have the same amount of vitamin. However, in the samples treated with the electric field causes a decrease in the vitamin from 3 minutes of treatment, (55.5%), the decrease continues at a speed of 0.21 min^{-1}, until reaching 1.5% ascorbic acid residual in 20 min. The effect of EF on this vitamin is similar to the one exerted on β-carotene standard, although more severe, indicating that vitamin C is more labile facing environmental agents. Figure 17 shows the corresponding chromatograms of ascorbic acid with and without the EF treatment.

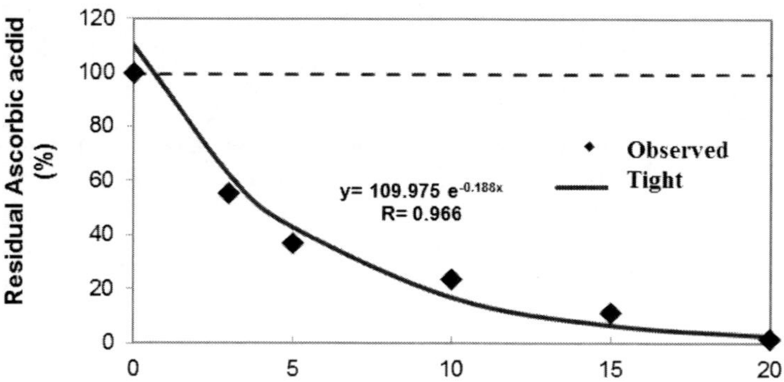

Figure 16. Residual ascorbic acid from samples treated with EF (245 nm).

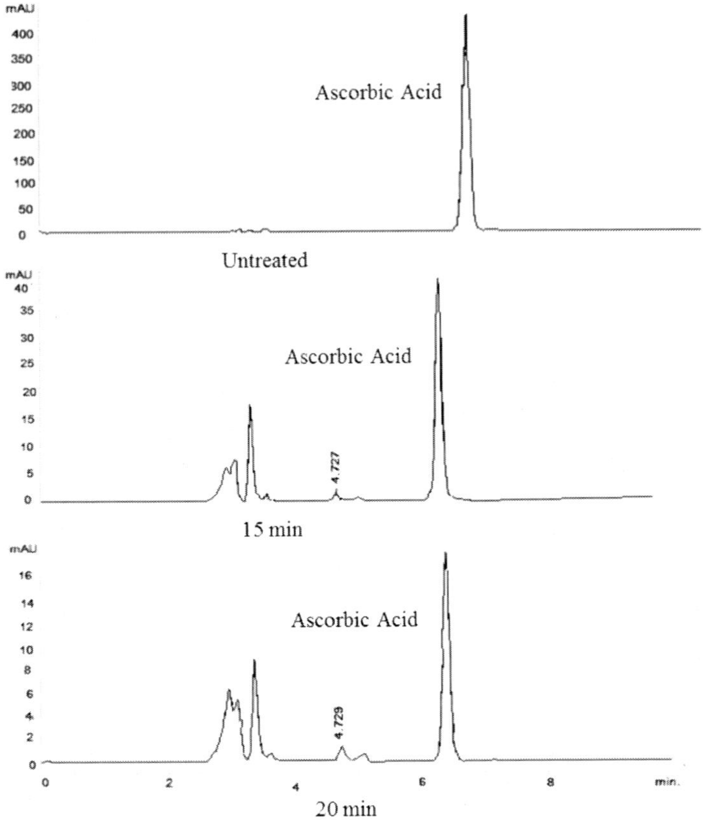

Figure 17. Chromatograms of L-ascorbic acid untreated and treated with EF.

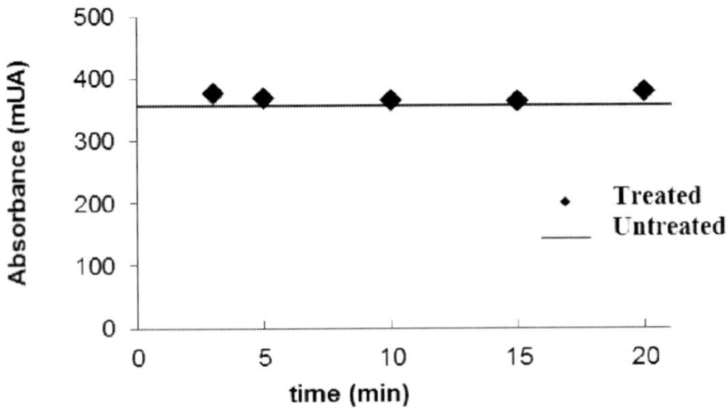

Figure 18. Residual ascorbic acid in avocado pastes treated with EF (245 nm).

The areas of the ascorbic acid peaks of the control are more significant than the treated samples as the treatment time increases, with a new peak appearing at 4.7 min. These results suggest that the electric field caused the degradation of ascorbic acid, resulting in the generation of a compound, which is more polar than the vitamin, based on the order of elution. These results confirm the decrease of ascorbic acid by the EF effect.

On the other hand, Figure 18 shows the readings of the area under the curve of ascorbic acid when it is in avocado paste treated with EF.

Once more, the concentration of this vitamin is not affected when it is in the avocado paste, because the avocado matrix is a very complex matrix that surrounds the products, or else the effect of the electric field is diluted between the vast amounts of molecules that are in the matrix.

CONCLUSION

The effect of EF on chemically pure β-carotene, analyzed with UV/visible spectroscopy, suggests the formation of geometric isomers with higher polarity as a function of the treatment times ($p < 0.05$), however, when β-carotene is immersed in the avocado matrix shows much fewer modifications.

The electric field causes the degradation of pure L-ascorbic acid as the treatment time increases, while, in the avocado matrix, the vitamin is not affected after the treatments. It seems that all the components of the avocado paste provide a protective effect to the β-carotene and ascorbic acid vitamins.

About α-tocopherol, no significant damage caused by EF was detected in all cases, neither in the standard nor in the avocado paste, ($p > 0.05$), this vitamin that works as an excellent antioxidant is stable to treatments with the electric field.

In general, it was found that in the avocado matrix, no differences were observed with the control ($p > 0.05$). L-ascorbic acid, in the avocado paste, no degradation was observed due to the application of EF. This technology did not have a significant effect on pure α-tocopherol nor on avocado paste ($p > 0.05$), so its application can be recommended as an avocado paste preservation technique since both sensory and nutritional characteristics are preserved. Besides, it is possible to use this technology to preserve foods rich in vitamins and functional fatty acids.

The treatment with EF is, therefore, a technology with broad perspectives for the conservation of avocado pulp.

The technology of the electric field, using the equipment designed in the CIBA-IPN, has demonstrated that it diminishes the activity of the polyphenol oxidase enzyme, which causes the darkening of the avocado pulp, and in turn, decreases the microbial contamination.

Although the effectiveness of other emerging technologies such as PEF, high pressures, or microwaves has already been demonstrated to diminish enzymatic activities and microbial populations, these types of equipment are costly, and some of them do generate increases in temperature. It is worth highlighting the importance of the electric field system used in the present work, which does not generate heat with their treatments, also, that its cost represents only between 20 to 30% of the cost of commercial PEF equipment.

There are no studies published on the stability of vitamins and fatty acids when the activity of polyphenol oxidase is reduced under some treatment. In conclusion, the results obtained here could be a basis for the adaptation of this technology in the food industry as a conservation method respectful with the environment, of low energy cost, and with the conservation of sensory and nutritional attributes, with a team designed by Mexican researchers.

References

"Official Method 963.22. Methyl esters of fatty acids in oil and fats gas chromatographic method," In *Association Official of Analytical Chemists*, by AOAC, 84-89. USA: AOAC Official Methods Anal, 2009.

Bora, S. P., N Narain, R. V. M. Roche, and M. P. Queiroz. "Caracterización de los aceites de la pulpa y semillas de aguacate" ["Characterization of the pulp and avocado seed oils"], *Grasas y Aceites*, 2001: 171-174.

Castorena García, José Hugo. *Estudio del efecto de los campos electromagneticos* [*Study of the effect of electromagnetic fields*], Tesis para grado de Doctor, Tlaxcala, México, 2013.

Davey, M., J. Keulemans, and R. Swennen. "Methods for the efficient quantification of fruit provitamin A contents," *Journal of Chromatography A*, 2006: 1136-1184.

Davies, M., J. Austin, and D. Partridge. "Vitamin C," In *Its Chemistry and Biochemistry*, by Austin J., Partridge D. Davies M. Londres: The Royal Society of Chemistry Paperbacks, 1991.

Doevenspeck, H. "Influencing cells and cell walls by electrostatic impulses," *Fleischwirtschft*, 1961: 968-987.

Elez-Martínez, Pedro, Robert Solvia-Fortuny, and Olga Martín -Belloso. "Oxidative rancidity in avocado puree as affected by α-tocopherol,". *Eur Food Res Technol*, 2007: 295-300.

Gaydou, E. M., Y. Lozano and J. Ratovohery. "Triglyceride and fatty acid compositions in the mesocarp of *Persea americana* during fruit development," *Phytochemistry*, 1987: 26(6), 1595-1597.

Giner, J., V. Gimeno, G. V. Barbosa-Cánovas, and O. Martín-Belloso. "Effects of pulsed electric fields processing on apple and pear polyphenol oxidases." *Food Science and Technology International*. 2001: (7) 339-347.

Hautfenne, A. "Applied Chemistry division Commission on Oils, Section III. Glycerines," In *International Union of Pure and Applied Chemistry,* by IUPAC, Fats and Derivatives 6th ed. Belgium: Université Catholique de Louvain-la-Neuve, 1990.

López-Malo, A., E. Palou, G. V. Barbosa-Cánovas, J. Welti-Chanes, and B. G. Swanson. "Polyphenoloxidase activity and color changes during storage of high hydrostatic pressure treated avocado puree," *Food Research International*, 1999: 31, 549-556.

Mexbest Safety and Quality Taste. https://www.mexbest.com/blog-el-aguacate-mexicano-en-el-super-bowl-25.html. January 14, 2019. (accessed February 10, 2019).

Ortiz, M. A., L. Dorantes, J. Galíndez, and RI. Guzmán. "Effect of different extraction methods on fatty acids, volatile compounds, and physical and chemical properties of avocado (*Persea americana* Mill) oil," *J. Agric. Food Chem.*, 2003: 51, 2216-2221.

Planeación Agricola Nacional. *Aguacate Mexicano* [*Mexican avocado*]. www.gob.mx/sagarpa, 2016.

Qiu, D., Z. Chen and H. Li. "Effect of heating on solid beta-carotene." *Food Chemistry*, 2009: 112: 344-349.

Requejo-Tapia, C. L. *International trends avocado and avocado oil production and seasonal variation of fatty acids in New Zeland-grown cv. Hass*, Thesis of Master. Science in Agribusiness, University New Zeland, 1999.

Rodriguez-Amaya, D, and E Azebedo-Meleiro. "Confirmation of the identity of the carotenoids of tropical fruits by HPLC-DAD and HPLC-MS." *Journal of Food Composition and Analysis*, 2004: 17: 385-396.

Rohit, Thirumdas, Sarangapani Chaitanya, and Annapure Uday S. "Cold Plasma: A novel Non-Thermal Technology for Food Processing," *Food Biophysics*, 2014.

SADER. *Boletin mensual de la producción de aguacate* [*Monthly bulletin of avocado production*]. México: Servicio de Información Agroalimentaria y Pesquera, 2019.

Sears, Zemansky. "Carga electrica y campo electrico" ["Electric charge and electric field"] In *Fsisica Universitaria*, by Zemansky Sears, 716. España: open source, 2012.

Soliva-Fortuny, Robert C., Pedro Elez-Martínez, Merce Sebastián-Calderö, and Olga Martín-Belloso. "Kinetics of polyphenol oxidase activity inhibition and browning," *Journal of Food Engineering*, 2002: 131-137.

Torregrosa, F., C. Cortés C, M. Esteve, and A. Frígola. "Effect of High-Intensity Pulsed Electric Fields Processing and Conventional Heat

Treatment on Orange–Carrot Juice Carotenoids," *J. Agric. Food Chem*, 2005: 53: 9519-9525.

Tovar, Ángel Miguel Ortega. "Valor Nutrimental de la Pulpa Fresca de Aguacate Hass" ["Nutritional Value of Fresh Hass Avocado Pulp"], *Proceedings V Worlda Avocado Congress.*, 2003: 741-748.

Ume, Roobab, Muhammad Aadil Rana, Muhammad Madni Ghulam, and El-Din Bekhit Alaa. "The Impact of Nonthermal Technologies on the Microbiological Quality of Juices: A Review," *Institute of Food Technologists*, 2018: 437-457.

Weemaes, C., L. Ludikhuyze, I. Van den Broeck, and M. Hendrickx. "High pressure inactivation of polyphenol oxidase," *J. Food Sci.*, 1998: 63: 873-877.

Yeom, H. W., C. B. Streaker, H. Q. Zhan, and D. B Min. "Effects on pulsed electric field on the quality of orange juice and comparison with heat pasteurization," *Journal of Agricultural and Food Chemistry*, 2000: 5107-5141.

In: Avocado: Consumption and Health ISBN: 978-1-53618-204-0
Editors: M. G. Melgar Lalanne et al.

Chapter 9

Antimicrobial Activity of Phenolic Compounds in Virgin Avocado Oil

Mónica Hernández[1], Gerardo Sánchez-González[2], Alan Javier Hernández-Álvarez[3] and Guiomar Melgar Lalanne[4,*]

[1]Facultad de Química Farmaceútica Biológica,
Universidad Veracruzana, Xalapa, Veracruz, Mexico
[2]Unidad de Servicios de Apoyo en Resolución Analítica,
Universidad Veracruzana, Xalapa, Veracruz, México
[3]School of Food Science and Nutrition,
University of Leeds, Leeds, UK
[4]Instituto de Ciencias Básicas, Universidad Veracruzana,
Xalapa, Veracruz, México

[*] Corresponding Author's Email: gmelgar@uv.mx

ABSTRACT

Cold-pressed avocado oil (AO) is an unrefined oil from the fruit of *Persea americana*, high in monounsaturated fatty acids and unsaponifiable lipids. Moreover, AO is recognized for its positive effects on the prevention of cardiovascular diseases and metabolic disorders. The high content in tocopherols, carotenoids, phenolic compounds, and chlorophylls make this oil a potential source of natural antimicrobial compounds. Therefore, the objective of this research was to explore the antimicrobial activity of the water-soluble fraction of AO against foodborne pathogens. Results showed high levels of inhibition (>50%) against gram-positive Listeria monocytogenes ATCC 20225 and Listeria innocua ATCC 33090. These findings demonstrated that the water extract of AO, rich in polyphenolic compounds, had inhibition on some pathogenic strains, suggesting its potential as a natural preservative.

Keywords: avocado oil, antimicrobial activity, phenolic compounds

1. INTRODUCTION

Avocado (*Persea americana* Mill) is an endogenous tree native to south-central Mexico, which belongs to the Lauraceae family. The tree is cultivated for its fruits. There are three known races and many differential commercial varieties of avocado. The Hass variety is the most cultivated worldwide (Ramírez-Gil et al., 2019), and Mexico is the principal global avocado producer and exporter. The country produced more than 2 million Tons in 2017 (Tan et al., 2017). Although the fresh fruit is still the most consumed form, the use of avocado oil in the cosmetic, pharmaceutical, and food processing industries is increasing during the last decades for its nutritional and nutraceutical properties, and its unique sensorial characteristics (Tan et al., 2017). Moreover, avocado leaves, seeds, and peels have been traditionally used for centuries as infusions for their therapeutic properties such as antimicrobial, anti-inflammatory, antioxidant, and anticancerogenic, between others, which have been attributed to the high presence of polyphenolic compounds, carotenoids, and tocopherols (Jiménez et al., 2020).

The avocado oil is mostly extracted by using organic solvents (mainly hexane) and/or elevated temperatures, which negatively affects its final quality and could be used mainly for cosmeceutical purposes since its significant free fatty acid presence makes the oil unviable for human consumption. To obtain the oil, the fruit is peeled and destoned. Then the pulp is mashed and eventually dried. Finally, the pulp is heated with hot water and chalk and/or NaCl and spun, and then extracted with organic solvents. The use of these stressful conditions implies the necessity of a final refining process, the oil deacidification obtained by the elimination of free fatty acids to blanch the chlorophylls and other degradation products, and the deodorization of the oil by eliminating undesirable volatile compounds (Costagli and Betti, 2015). The cold-pressed avocado oil (virgin avocado oil) increases the final quality of the oil and maintains the chemical and sensorial characteristics of the fresh fruit. The oil is then extracted at low temperatures (<50°C) by using mechanical methods (pressing), and the final product is not refined. The extraction allows preserving the color pigment and heat-labile bioactive components as well as the natural aroma and flavor. Moreover, virgin avocado oil is considered a functional oil due to its elevated level of monosaturated fatty acids, phytosterols, antioxidants, carotenoids, and tocopherols (Tan, 2019).

The saponifiable fraction of the avocado oil is mainly composed of fatty acids associated with glycerol. This fraction is like olive oil. Therefore, nearly 70% of the oil is composed of monounsaturated fatty acids (oleic and palmitoleic acids), 17% of polyunsaturated (linoleic and linolenic acids), and 14% of saturated fatty acids (palmitic and stearic acids). This composition made this oil highly stable against heat (more than 176°C) and with a smoke point higher than 250°C, ideal for cooking and frying (Forero-Doria et al., 2017).

The unsaponifiable fraction of the avocado oil is of significant interest in its content in carotenoids, chlorophylls, tocopherols, and phenolic compounds, which has shown antimicrobial, antioxidant, and anti-inflammatory properties (Zhang et al., 2013). The pulp of avocado fruits contains between 60 to 168 mg of gallic acid equivalents per g (Krumreich et al., 2018) lower than the seed, peel, and leaves (Jiménez et al., 2020).

This composition also depends on the avocado variety analyzed, the extraction technology used, and the ripening stage of the fruit (Santana et al., 2019). Then, the presence of peels in the oil increases the presence of this unsaponifiable fraction, and seven phenolic compounds have been identified (gallic acid, 3,4 di-OH-*p*-phenylacetic acid, *p*-OH-benzoic acid, vanillic acid, *p*-coumaric acid, ferulic acid, and quercetin). Moreover, in general, quercetin is present in all unripe and ripe samples peeled and unpeeled, and gallic acid is the most common found in ripe and unpeeled samples (Rodriguez-Carpena et al., 2011; Santana et al., 2019). The cold pressing (60°C) improves the presence of polyphenol compounds in avocado oil when is compared with solvent extraction in the same conditions, and it is related to the antioxidant activity of the oil (Krumreich et al., 2018).

The antimicrobial action of some oils is widely described. Therefore, essential oils from spices and aromatic herbs are routinely used in the food industry as an additive for their sensorial properties and their preservative properties. Their antimicrobial action has been described both against Gram-positive and Gram-negative enteropathogenic strains like E. coli, S. aureus, *S. typhi, P. aeruginosa*, and *E. faecalis*, being the main responsible for this activity terpenes and pyrenes (Valdivieso-Ugarte et al., 2019). Edible oils have also been described as potential sources of antimicrobial compounds due to their high content in phenolic acids and flavonoids. Edible oils such as sunflower, safflower, canola, soybean, and cotton from Japan showed antimicrobial activity against S. aureus and E. coli (Xuan et al., 2018). Moreover, cold-pressed oils usually have more polyphenolic compounds in their composition and are potentially useful antimicrobial sources (Boskou, 2017). Oleic acid exhibits a variety of biological activity including antimicrobial against some fungus, yeast (*A. niger, F. heterosporous, P. valentinum, S. cerevisiae, H. anomala*, and *S. pararoseus*), and bacteria (*B. subtilis, S. aureus,* and *E. coli*) with a MIC between 125 to 500 μg/mL (Ozsen Batur and Kiran, 2019).

The water fraction of avocado oil, and in general from most edible oils, is mainly composed of polyphenols, carotenoids, and tocopherols. In this sense, there is enough evidence that suggests the beneficial effect of the phenolic fraction on reducing the risk of cardiovascular diseases by reducing the cell oxidative and inflammatory stress (Dreher and Davenport, 2013). This direct health benefit is causally related to the high antioxidant activity of the polyphenols. Besides the health benefits that the presence of antioxidants provides, this fraction is also responsible for the high shelf life and storage stability of the oil. Moreover, the phenolic acids also work as food preservatives, since their antimicrobial potential has been related to their saturated chain length, position, and number of substitutions in the core benzene ring. Like other weak organic acids, phenolics can acidify the cytoplasm leading to cell death.

Consequently, the pKa and hydrophobicity determine the solubility of the phenolic compound in the microbial membrane and are responsible for the antimicrobial activity. In this sense, the pH of the medium plays an important role, since it can change the saturation of the side chain, and so, the antimicrobial capacity of the phenolic acid (Kumar and Goel, 2019). This characteristic makes polyphenols (special phenolic acids) a suitable alternative to increase the shelf life of food products by reducing the fat oxidation, and they may act as antimicrobials against pathogenic microorganisms.

The antimicrobial action of the phenolic compounds is due to the combination of diverse mechanisms. Hydroxyls interact with the bacterial cell membrane by hydrogen bonding and can cause the disruption of the cell membrane structure or may depolarize the membrane by reducing the pH gradient across the membrane and the level of the ATP pool. The presence of alkyl groups in the aromatic nucleus generates phenoxyl radicals, which can alter the hydrophobicity of the cell membrane. Phenolic compounds showed their antimicrobial action at a cellular level in diverse ways:

(1) by the induction of the leakage of the cellular content;
(2) by interrupting the DNA, RNA, and protein synthesis and function;
(3) by the induction of the acidification of the cytoplasm and the coagulation of some cytoplasmic components;
(4) by the disruption of the membrane function and structure;
(5) by the interference with intermediary metabolism, and
(6) by the modification of the membrane potential (Bourabab Chibane et al., 2019).

The most qualified cold-pressed edible oil is the extra-virgin olive oil, whose antimicrobial activity has been widely described and related to the hydrophilic fraction, mostly with the phenolic compounds. This fraction has been reported as effective against a broad spectrum of foodborne pathogens. Furthermore, this has been used to prevent and avoid the spoilage in foods contaminated with *L. monocytogenes* (Guo et al., 2019). The Algerian oil varieties showed antimicrobial activity mainly against *S. aureus* and *B. subtilis*. However, the antimicrobial activity depended on the oil variety analyzed (Laincer et al., 2014). Nazzario et al., 2019 studied the polyphenolic fraction in olive oil and confirmed the antibacterial properties against foodborne pathogens. The composition of the phenolic fraction varies with the cultivar, ripeness, weather conditions, and oil extraction process. However, the main responsible for the antimicrobial activity may be the unsaponifiable fraction, which includes the presence of squalene, beta-sitosterol, and phenols (Hatami et al., 2016).

Small peptides present in the avocado pulp have demonstrated antimicrobial activity against *E. coli* and *S. aureus (*Guzmán Rodríguez et al., 2013*)*. Similarly, methanolic extracts from pulp have shown antimycobacterial activity against Mycobacterium tuberculosis. Organic extract from seeds inhibited the growth of strains of Candida spp. *Cryptococcus neoformans, and Malassezia pachydearmatis, S. aureus, S. pyogenes, C. ulcerans, C. albicans, E. coli,* and *S. typhi* (Jimenez et al., 2020). However, the specific polyphenols responsible for this activity have not been determined.

Subsequently, to our knowledge, the antimicrobial activity of the water fraction of cold-pressed avocado oil, rich in polyphenols, has not been reported, although some published work about the antibacterial activity of avocado seeds, peels, and fruit has been carried out. Therefore, this research aimed to explore the antimicrobial activity of the water extract fraction of virgin avocado oil extracted by cold pressing.

2. Material and Methods

2.1. Biological Material

The press-cold extracted virgin avocado oil used in the present research was purchased in a local supermarket (Ahuacatlan, DIRICOM, SA de CV, Mexico). The oil did not declare added preservatives. The total polyphenols extracted from virgin avocado oil was tested against six pathogenic strains: *Escherichia coli* ATCC 4385, *Salmonella enterica subsp. enterica* (ex Kauffmann and Edwards) Le Minor and Popoff serovar Typhimurium ATCC 14028, *Listeria monocytogenes* ATCC 20225, *Listeria innocua* ATCC 33090, *Staphylococcus aureus* ATCC 25923, and *S. aureus* FRI 481.

2.2. Reagents and Culture Media

Salts, acids, alkalis, and chemical indicators used in this work were ACS-grade. The rest of the reagents were purchased from Sigma-Aldrich (St. Louis, MO, USA). Mueller-Hinton broth (2 g/L beef extract; 17.5 g/L acid digestion of casein; 1.5 g/L starch) was purchased from Difco (Franklin Lakes, NJ, USA).

2.3. Determination of Total Polyphenols in Avocado Oil

The phenolic compounds present in the avocado oil were determined by the Folin test (Cerecedo-Cruz et al., 2018). Briefly, The Folin-Ciocalteu reagent was diluted in deionized water 1:10 (w/w) before use, and each avocado oil sample was diluted 1:5 (w/w) in ethanol. Then, 0.1 mL of the dilution was mixed with 0.75 mL of the Folin-Ciocalteau reagent. The solution was incubated at room temperature for 5 min, and 0.75 mL of sodium bicarbonate (60 g/L) was added and mixed. The sample was then incubated at 25°C for 90 min and filtered using a 0.45 μm syringe filter (Whatman, International Ltd., Maidstone, Kent, UK). Finally, the Absorbance was measured at 750 nm, and gallic acid was used as standard. Results were expressed as μg gallic acid equivalents/mL of oil.

2.4. Extraction of the Hydrophilic Fraction of Avocado Oil

The extraction of the total hydrophilic fraction of avocado oil was performed by following the method of Nazzaro et al., 2019. Briefly, 10 mL of avocado oil was mixed with 10 mL of PBS buffer (8.0 g/L NaCl, 0.2 g/L KCl, 0.541 g/L KH2PO4, 0.854 g/L, Na2HPO4; pH 7.0) and μL of Polyethylene glycol sorbitan monolaurate (Tween 20, Sigma-Aldrich, USA) under sterility. Then, the mixture was mixed using a vortex (5 min) and in an orbital shaker (200 rpm, 1h). After that, the mixture was centrifuged (2000 g, 3 min, 20°C), and the oleic phase was kindly separated with a Pasteur pipette. The aqueous phase was centrifuged for a second time, and the oily phase was then separated. The final aqueous phase was used for the antimicrobial activity determination assay.

2.5. Antibacterial Activity

Culture Conditions of Pathogenic Strains

All the pathogenic strains used for the experiments were stored frozen at – 20°C in Mueller-Hinton broth with 30% glycerol (v/v) until their use.

Moreover, previously to the experiment, strains were activated at least twice in Mueller-Hinton broth at 37°C, 24 h under static and aerobic conditions.

Determination of the Antibacterial Activity

For the well-diffusion agar assay (Dussault et al., 2014), the target strains were spread onto Petri dishes with soft Mueller-Hinton agar (1%, w/v). Therefore, 6 mm diameters disks were impregnated with five μL of avocado oil for 20 min at room temperature. Each Petri dish was inoculated with 100 μL of each bacterial suspension. When the bacterial suspension was wholly absorbed in the Petri dish, the disk previously treated was punched and then incubated at 37°C for 24 h. Chloramphenicol was used as negative growth control. The diameter of the inhibition halo was measured by triplicate and reported.

The microdilution method was performed, according to Watkins et al., 2013. Previously to the experiments, the initial bacteria concentration was adjusted to A600nm = 1.0 (approx. 107CFU/mL) in Mueller Hinton broth. The five pathogenic strains were screened for the antimicrobial activity in 96-well microtitre plates. Each sample well was then filled with 90 μL of Mueller Hinton broth, 100 μL of water extracted, and ten μL of the bacteria inoculum. As growth control, wells with 90 μL of Mueller Hinton broth, 100 μL of (PBS+Tween 20), and ten μL of the bacteria inoculum were used. Chloramphenicol was used as a positive control for the antimicrobial activity. Each experiment was run four times in two independent assays (n = 4). Bacterial growth was determined by the lecture of the Absorbance at 800 nm at 0 h and 24 h after incubation at 37°C. The inhibition of bacterial growth was determined as the difference between the final and initial Absorbance.

2.6. Statistical Analysis

Data were expressed as mean ± standard derivation of quadruplicate measurements by using the Sigma Plot 12.5 software (Systat Software,

USA) for calculations. The significant level at 1% between means was determined by one-way analysis of variance (ANOVA).

3. Results and Discussion

Oils extracted from plants have been commonly used for the treatment of different diseases, and nowadays, there is a growing interest in finding the natural compounds responsible for these traditional uses. Hence, the industry is trying to develop new preservatives that can be an alternative to synthetic antimicrobial agents with less secondary environmental impacts and reduced adverse health effects. Essential oils have been widely studied in this sense. They are mostly extracted with organic solvents (mainly ethanol, hexane, and methanol) or water extracted through steam dragging at high temperatures, which negatively affect their final quality. Cold pressing extraction gives final products with a higher presence of bioactive metabolites from the sample extracted. Currently, it is common to obtain virgin olive oil and avocado oil through this process, although recently, it has been proposed to extract oils from other high-value seeds (Grosshagauer et al., 2019). The food use of plant extracts is promising for the industry since most of them possess the GRAS (generally recognized as safe) status, which is mostly added to beverages, fermented milks, and smoothies (Bourab et al., 2019).

Antimicrobial and antioxidant activities are firmly related, since antioxidant metabolites can act against the excess of reactive oxygen species, which lead to lipid peroxidation, and then act over the cell lipid membrane through different mechanisms as previously mentioned (Bourab et al., 2019; Valdivieso-Ugarte et al., 2019). Virgin olive oils have anti-atherosclerotic potential, favoring endothelial function and preserving blood pressure, maintaining lipoprotein functionality, exerting anti-inflammatory and antioxidant effects, and modulating gene expression in several tissues to maintain a proper homeostasis (Gaforio et al., 2019). Many of these effects are related to the polyphenols, which act over the cells. However, the biochemical and biological mechanisms remain still poorly documented, and it is necessary to understand the pharmacokinetics and pharmacodynamics of the polyphenols to understand the effective dose in humans. The dose-dependence of these effects and clinical trials are needed to overcome the limits of their benefits. (Rigacci and Stefani, 2016; Gorzynik-Debicka et al., 2018).

Few researches have noticed the antimicrobial activity of avocado fruit, mostly from peels and seeds, which are rich in polyphenolic compounds (Bhuyan et al., 2019). However, the avocado oil has been poorly studied for its antimicrobial activity even when it is considered a mid-range source of polyphenols with a similar lipid profile like virgin olive oil.

Results showed a total polyphenol content in the cold-pressed virgin avocado oil of 40.30 ± 3.39 mg gallic acid equivalents/kg, slightly less than previously found by Flores et al., 2014 in Chilean avocado oil (between 42.6 to 56.9 mg/kg), but higher than that found by Santos et al., 2018 (0.37 mg gallic acid equivalent/100 mL). The differences could be related to the cultivar, the ripening stage of the fruit, the growing conditions, and other environmental conditions. The technology used also can influence the polyphenol extraction from the fruit. Thus, the preconditioning of the fruit, the pressing conditions (with or without the seed for example), and the extraction temperature impact the oil's final composition. Furthermore, the storage conditions (temperature, time, light, and exposition) should also be considered (Rigacci and Stefani, 2016).

Avocado is considered a mid-range fruit for its phenolic levels (2 to 4.6 mg gallic acid equivalents/g of fruit) and has a great fruit lipophilic antioxidant capacity mainly due to this fraction (Deher and Davenport, 2013). Although avocado oil has not been deeply analyzed for its beneficial properties in human health, there is much information about virgin olive oil in which polyphenol composition is considered close to the virgin avocado oil. In olive oil, the polyphenols have shown antioxidant properties, which may help to maintain healthy LDL cholesterol levels and reduce lipid oxidation. Moreover, these compounds act in lipid metabolism when absorbed in the small intestine and metabolized. They act by stabilizing the olive oil by protecting the oil against oxidation, and this contributes to a good HDL cholesterol level and may help to protect LDL cholesterol and lipid oxidation (Tetens, 2011).

The antimicrobial activity of some common edible oils has been analyzed by Medina et al., 2007, who found that refined oils like sunflower and corn oil did not show antimicrobial activity, probably due to the loss of polyphenols during the refining process. Moreover, only virgin olive oils samples (extracted by cold pressing) showed antimicrobial activity, which may suggest that this activity is due to compounds different from the fraction of the fatty acids that are present in oils, mostly in the water fraction, like polyphenols.

The antimicrobial activity of avocado extracts might be influenced by (a) the variety of avocado; (b) the part investigated (leaves, peels pulp, seeds, oily fraction); (c) the extraction conditions; and (d) the microbial pathogenic species analyzed (Bhuyan et al., 2019). In general, the highest antimicrobial activity has been found in polar extracts (ethanolic, aqueous, and methanolic) (Melgar-Lalanne et al., 2017). However, other authors described no antimicrobial activity from avocado extracts in boiling distilled water, probably because of the oxidation of the antioxidant compound with the high temperature and the contact with water (Calderon-Oliver et al., 2016).

The antimicrobial activity of avocado oil in agar diffusion assay is shown in Figure 1, and the experimental results obtained from the antimicrobial activity of the water fraction of virgin avocado oil by the microdilution test is shown in Figure 2. Results showed that avocado oil had good antimicrobial activity against Gram-positive strains like *L. monocytogenes* and *S. aureus,* like in previous reports (Santos et al., 2018). The antimicrobial activity resulted significantly higher ($p > 0.05$) in all the strains tested in the water fraction analyzed with the microdilution assay than in the total oil analyzed with the agar diffusion assay. These results reinforce the hypothesis mentioned above of other authors that the antimicrobial activity is related to the antioxidant activity and, thus, with the total polyphenol content (Medina et al., 2007). However, other authors had not found any antimicrobial activity against Gram-positive strains like *S. aureus, L. monocytogenes* and *B. aureus*, and medium activity against

Gram-negative strains like *S. Typhimurium* and *S. enteritidis* (Santos et al., 2018), when cold-pressed Brazilian avocado oil was analyzed by the agar diffusion assay. Differences in both studies may be due to environmental, technological, and storage differences in the cold avocado oils analyzed.

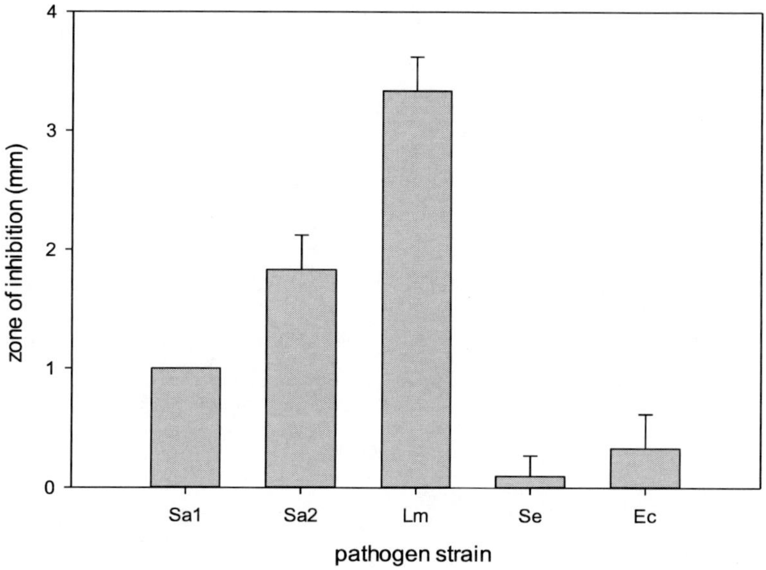

Figure 1. Size of the inhibition halo (mm) of the water fraction isolated from virgin avocado oil, where Ea is *Escherichia coli* ATCC 4385, Se is *Salmonella enterica subsp. enterica (ex Kauffmann and Edwards) Le Minor and Popoff serovar Typhimurium* ATCC 14028, Lm is *Listeria monocytogenes* ATCC 20225, Sa1 is *Staphylococcus aureus* ATCC 25923, and Sa2 is *S. aureus* FRI 481. $n = 8$. $p < 0.01$. The same subscript letter means no significant difference ($p > 0.01$). Each error bar represents the mean and standard derivation ($n = 6$).

Rodríguez-Carpena et al., 2011 analyzed extracts of avocado peel, pulp, and seeds in ethyl acetate, 70% of acetone, and 70% of methanol and found antimicrobial activity against Gram-positive bacteria, *B. cereus* and *L. monocytogenes*, more than against Gram-negative (being *E. coli* and *Pseudomonas* spp. the most sensitive). The avocado extract also acted against *Y. lipolytica*, but not against *A. niger*. Moreover, acetogenins from the avocado lipid fraction show a wide range of biological activities including antimicrobial, antioxidant, antiplatelet, and antithrombotic, and had shown antilisterial activity against *L. monocytogenes* both *in vitro* and in a refrigerated food matrix (Salinas-Salazar et al., 2016; Villareal-Lara et al., 2019) and against *Clostridium sporogenes* (Pacheco et al., 2017).

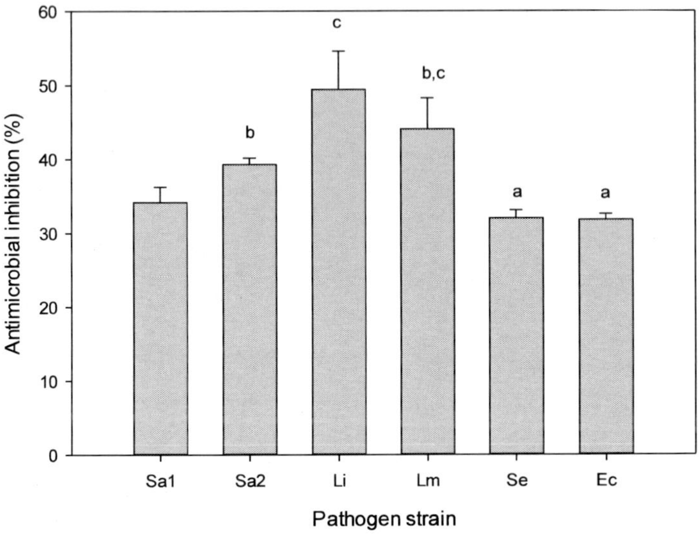

Figure 2. Antimicrobial inhibition (%) of the water fraction isolated from virgin avocado oil, where Ea is *Escherichia coli* ATCC 4385, Se is *Salmonella enterica subsp. enterica (ex Kauffmann and Edwards) Le Minor and Popoff serovar Typhimurium* ATCC 14028, Lm is *Listeria monocytogenes* ATCC 20225, Li is *Listeria innocua* ATCC 33090, Sa1 is *Staphylococcus aureus* ATCC 25923, and Sa2 is *S. aureus* FRI 481. $n = 8$. $p < 0.01$. The same subscript letter means no significant difference ($p > 0.01$). Each error bar represents the mean and standard derivation ($n = 6$).

Conclusion

In this research, the water extract from avocado oil rich in polyphenols confirms its antibacterial properties, mostly against Gram-positive strains. However, more pathogenic strains and species should be analyzed in vitro. Future researches should be focused on the identification of the individual molecules responsible for this activity to better design a new natural preservative based on the avocado water fraction. Moreover, complementary studies about this ideal formulation in different food matrixes are also recommended by using different storage conditions.

REFERENCES

Batur, Ö. Ö., Atlı, Ö., and Kıran, İ. (2019). Biotransformation of oleic acid and antimicrobial and anticancer activities of its biotransformatıon extracts. *Bulgarian Chemical Communications,* 51 (2), 200-205.

Bhuyan, D. J., Alsherbiny, M. A., Perera, S., Low, M., Basu, A., Devi, O. A., Barooah, M. S., Li, C. G., and Papoutsis, K. (2019). The Odyssey of Bioactive Compounds in Avocado (*Persea americana*) and Their Health Benefits. *Antioxidants,* 8(10), 426. https://doi.org/10.3390/antiox8100426.

Boskou, D. (2017). Edible cold-pressed oils and their biologically active components. *Journal of Experimental Food Chemistry,* 3(1), e108.

Bouarab Chibane, L., Degraeve, P., Ferhout, H., Bouajila, J., and Oulahal, N. (2019). Plant antimicrobial polyphenols as potential natural food preservatives. *Journal of the Science of Food and Agriculture*, 99(4), 1457–1474. https://doi.org/10.1002/jsfa.9357.

Calderón-Oliver, M., Escalona-Buendía, H. B., Medina-Campos, O. N., Pedraza-Chaverri, J., Pedroza-Islas, R., & Ponce-Alquicira, E. (2016). Optimization of the antioxidant and antimicrobial response of the combined effect of nisin and avocado byproducts. *LWT - Food Science and Technology,* 65, 46–52. https://doi.org/https://doi.org/10.1016/j.lwt.2015.07.048.

Cerecedo-Cruz, L., Azuara-Nieto, E., Hernández-Álvarez, A., González-González, C., and Melgar-Lalanne, G. (2018). Evaluation of the oxidative stability of Chipotle chili (Capsicum annuum L.) oleoresins in avocado oil. *Grasas y Aceites,* 69(1), e240. http://dx.doi.org/10.3989/gya.0884171.

Costagli, G., and Betti, M. (2015). Avocado oil extraction processes: method for cold-pressed high-quality edible oil production versus traditional production. *Journal of Agricultural Engineering,* 46(3), 115-122. https://doi.org/10.4081/jae.2015.467.

Dreher, M. L., & Davenport, A. J. (2013). Hass Avocado Composition and Potential Health Effects. *Critical Reviews in Food Science and Nutrition,* 53(7), 738–750. https://doi.org/10.1080/10408398.2011.556759.

Dreher, M. L., & Davenport, A. J. (2013). Hass Avocado Composition and Potential Health Effects. *Critical Reviews in Food Science and Nutrition,* 53(7), 738–750. https://doi.org/10.1080/10408398.2011.556759.

Dussault, D., Vu, K. D., & Lacroix, M. (2014). In vitro evaluation of antimicrobial activities of various commercial essential oils, oleoresin and pure compounds against food pathogens and application in ham. *Meat Science,* 96(1), 514–520. https://doi.org/https://doi.org/10.1016/j.meatsci.2013.08.015.

Flores, M. A., Perez-Camino, M. D. C., and Troca, J. (2014). Preliminary studies on composition, quality and oxidative stability of commercial avocado oil produced in Chile. *Journal of Food Science and Engineering,* 4(1), 21.

Forero-Doria, O., García, M. F., Vergara, C. E., and Guzman, L. (2017). Thermal analysis and antioxidant activity of oil extracted from pulp of ripe avocados. *Journal of Thermal Analysis and Calorimetry,* 130(2), 959–966. https://doi.org/10.1007/s10973-017-6488-9.

Gaforio, J. J., Visioli, F., Alarcón-de-la-Lastra, C., Castañer, O., Delgado-Rodríguez, M., Fitó, M., Hernández, A. F., Huertas, J. R., Martínez-González, M. A., Menendez, J. A., Osada, J., Papadaki, A., Parrón, T., Pereira, J. E., Rosillo, M. A., Sánchez-Quesada, C., Schwingshackl, L., Toledo, E., and Tsatsakis, A. M. (2019). Virgin Olive Oil and Health: *Summary of the III International Conference on Virgin Olive Oil and Health Consensus Report, JAEN* (Spain) 2018. *Nutrients,* 11, 2039.

Gorzynik-Debicka, M., Przychodzen, P., Cappello, F., Kuban-Jankowska, A., Marino Gammazza, A., Knap, N., Wozniak, M., and Gorska-Ponikowska, M. (2018). Potential Health Benefits of Olive Oil and

Plant Polyphenols. *International Journal of Molecular Sciences,* 19(3), 686. https://doi.org/10.3390/ijms19030686.

Grosshagauer, S., Steinschaden, R., and Pignitter, M. (2019). Strategies to increase the oxidative stability of cold pressed oils. *LWT,* 106, 72–77. https://doi.org/https://doi.org/10.1016/j.lwt.2019.02.046.

Guo, L., Sun, Q., Gong, S., Bi, X., Jiang, W., Xue, W., and Fei, P. (2019). Antimicrobial Activity and Action Approach of the Olive Oil Polyphenol Extract against Listeria monocytogenes. *Frontiers in Microbiology,* 10, 1586. https://www.frontiersin.org/article/10.3389/fmicb.2019.01586.

Guzmán-Rodríguez, J. J., López-Gómez, R., Suárez-Rodríguez, L. M., Salgado-Garciglia, R., Rodríguez-Zapata, L. C., Ochoa-Zarzosa, A., & López-Meza, J. E. (2013). Antibacterial Activity of Defensin PaDef from Avocado Fruit (*Persea americana* var. drymifolia) Expressed in Endothelial Cells against *Escherichia coli* and *Staphylococcus aureus. BioMed Research International,* 986273. https://doi.org/10.1155/2013/986273.

Hatami, S., Mohamadi Sani, A. and Yavarmanesh, M. (2016), "Chemical composition and antibacterial activity of organic extra virgin olive oil from Iran," *Nutrition and Food Science,* 46 (3), 388-395. https://doi.org/10.1108/NFS-01-2016-0010.

Jimenez, P., Garcia, P., Quitral, V., Vasquez, K., Parra-Ruiz, C., Reyes-Farias, M., ... Soto-Covasich, J. (2020). Pulp, Leaf, Peel and Seed of Avocado Fruit: A Review of Bioactive Compounds and Healthy Benefits. *Food Reviews International,* 1–37. https://doi.org/10.1080/87559129.2020.1717520.

Krumreich, F. D., Borges, C. D., Mendonça, C. R. B., Jansen-Alves, C., and Zambiazi, R. C. (2018). Bioactive compounds and quality parameters of avocado oil obtained by different processes. *Food Chemistry,* 257, 376–381. https://doi.org/https://doi.org/10.1016/j.foodchem.2018.03.048.

Kumar, N., and Goel, N. (2019). Phenolic acids: Natural versatile molecules with promising therapeutic applications. *Biotechnology*

Reports, 24, e00370. https://doi.org/https://doi.org/10.1016/j.btre.2019.e00370.

Laincer, F., Laribi, R., Tamendjari, A., Arrar, L., Rovellini, P., and Venturini, S. (2014). Olive oils from Algeria: Phenolic compounds, antioxidant and antibacterial activities. *Grasas y Aceites,* 65(1), e001. http://dx.doi.org/10.3989/gya.035713.

Medina, E., Romero, C., Brenes, M., and de CASTRO, A. (2007). Antimicrobial Activity of Olive Oil, Vinegar, and Various Beverages against Foodborne Pathogens. *Journal of Food Protection,* 70(5), 1194–1199. https://doi.org/10.4315/0362-028X-70.5.1194.

Melgar-Lalanne, G., Hernández-Álvarez, A. J., Jiménez-Fernández, M., & Azuara, E. (2017). Oleoresins from Capsicum spp.: Extraction Methods and Bioactivity. *Food and Bioprocess Technology,* 10(1), 51–76. https://doi.org/10.1007/s11947-016-1793-z.

Nazzaro, F., Fratianni, F., Cozzolino, R., Martignetti, A., Malorni, L., De Feo, V., Cruz, A. G., and d'Acierno, A. (2019). Antibacterial Activity of Three Extra Virgin Olive Oils of the Campania Region, Southern Italy, Related to Their Polyphenol Content and Composition. *Microorganisms,* 7, 321.

Pacheco, A., Rodríguez-Sánchez, D. G., Villarreal-Lara, R., Navarro-Silva, J. M., Senés-Guerrero, C., and Hernández-Brenes, C. (2017). Stability of the antimicrobial activity of acetogenins from avocado seed, under common food processing conditions, against Clostridium sporogenes vegetative cell growth and endospore germination. *International Journal of Food Science and Technology,* 52(11), 2311–2323. https://doi.org/10.1111/ijfs.13513.

Ramírez-Gil, J. G., Cobos, M. E., Jiménez-García, D., Morales-Osorio, J. G., and Peterson, A. T. (2019). Current and potential future distributions of Hass avocados in the face of climate change across the Americas. *Crop and Pasture Science,* 70(8), 694–708. https://doi.org/10.1071/CP19094.

Rigacci, S., and Stefani, M. (2016). Nutraceutical Properties of Olive Oil Polyphenols. An Itinerary from Cultured Cells through Animal Models to Humans. *International journal of molecular sciences,* 17(6), 843. https://doi.org/10.3390/ijms17060843.

Rodríguez-Carpena, J. G., Morcuende, D., and Estévez, M. (2011). Avocado by-products as inhibitors of color deterioration and lipid and protein oxidation in raw porcine patties subjected to chilled storage. *Meat Science,* 89(2), 166–173. https://doi.org/https://doi.org/10.1016/j.meatsci.2011.04.013.

Salinas-Salazar, C., Hernández-Brenes, C., Rodríguez-Sánchez, D. G., Castillo, E. C., Navarro-Silva, J. M., and Pacheco, A. (2017). Inhibitory Activity of Avocado Seed Fatty Acid Derivatives (Acetogenins) Against Listeria Monocytogenes. *Journal of Food Science,* 82(1), 134–144. https://doi.org/10.1111/1750-3841.13553.

Santana, I., Castelo-Branco, V. N., Guimarães, B. M., Silva, L. de O., Peixoto, V. O. D. S., Cabral, L. M. C., … Torres, A. G. (2019). Hass avocado (*Persea americana* Mill.) oil enriched in phenolic compounds and tocopherols by expeller-pressing the unpeeled microwave dried fruit. *Food Chemistry,* 286, 354–361. https://doi.org/https://doi.org/10.1016/j.foodchem.2019.02.014.

Santos, J. S., Escher, G. B., da Silva Pereira, J. M., Marinho, M. T., Prado-Silva, L. do, Sant'Ana, A. S., … Granato, D. (2018). 1H NMR combined with chemometrics tools for rapid characterization of edible oils and their biological properties. *Industrial Crops and Products,* 116, 191–200. https://doi.org/https://doi.org/10.1016/j.indcrop.2018.02.063.

Tan, C. X. (2019). Virgin avocado oil: An emerging source of functional fruit oil. *Journal of Functional Foods,* 54, 381–392. https://doi.org/https://doi.org/10.1016/j.jff.2018.12.031.

Tan, C. X., Tan, S. S., and Tan, S. T. (2017). Influence of Geographical Origins on the Physicochemical Properties of Hass Avocado Oil. *Journal of the American Oil Chemists' Society,* 94(12), 1431–1437. https://doi.org/10.1007/s11746-017-3042-7.

Tetens, I. (2011). EFSA Panel on Dietetic Products, Nutrition and Allergies (NDA); *Scientific Opinion on the substantiation of a health claim related to glucosamine and maintenance of joints pursuant to Article 13 (5) of Regulation (EC) No 1924/2006.*

Valdivieso-Ugarte, M., Gomez-Llorente, C., Plaza-Díaz, J.and Gil, Á. (2019). Antimicrobial, Antioxidant, and Immunomodulatory Properties of Essential Oils: A Systematic Review. *Nutrients,* 11, 2786. https://doi.org/10.3390/nu11112786.

Villarreal-Lara, R., Rodríguez-Sánchez, D. G., Díaz De La Garza, R. I., García-Cruz, M. I., Castillo, A., Pacheco, A., and Hernández-Brenes, C. (2019). Purified avocado seed acetogenins: Antimicrobial spectrum and complete inhibition of Listeria monocytogenes in a refrigerated food matrix. *CyTA - Journal of Food,* 17(1), 228–239. https://doi.org/10.1080/19476337.2019.1575908.

Xuan, T. D., Gangqiang, G., Minh, T. N., Quy, T. N., and Khanh, T. D. (2018). An Overview of Chemical Profiles, Antioxidant and Antimicrobial Activities of Commercial Vegetable Edible Oils Marketed in Japan. *Foods,* 7(2), 21. https://doi.org/10.3390/foods7020021.

Zhang, Z., Huber, D. J., and Rao, J. (2013). Antioxidant systems of ripening avocado (*Persea americana* Mill.) fruit following treatment at the preclimacteric stage with aqueous 1-methylcyclopropene. *Postharvest Biology and Technology,* 76, 58–64. https://doi.org/https://doi.org/10.1016/j.postharvbio.2012.09.003.

In: Avocado: Consumption and Health ISBN: 978-1-53618-204-0
Editors: M. G. Melgar Lalanne et al. © 2020 Nova Science Publishers, Inc.

Chapter 10

FATTY ACID PROFILE OF OIL FROM HASS AVOCADO SEED (*PERSEA AMERICANA* MILL.) EXTRACTED BY THREE DIFFERENT METHODS

Alma Yadira Salazar-Govea[1,*], Gabriel Arellanes-Jiménez[1], Elvira Ríos-Leal[2] and Mirna Patricia Santiago-Gómez[1,†]

[1]Instituto de Agroindustrias, Universidad Tecnológica de la Mixteca, Huajuapan de León, Oaxaca, México
[2]Departamento de Biotecnología y Bioingeniería del Centro de Investigación y de Estudios Avanzados del Instituto Politécnico Nacional, Ciudad de México, México

* Corresponding Author's Email: almasalazar@mixteco.utm.mx.
† Corresponding Author's Email: patsanmx@mixteco.utm.mx.

ABSTRACT

This chapter addresses the yield and fatty acid composition of oil extracted from the seeds of Hass avocado fruit (*Persea americana* Mill.) grown in Oaxaca, Mexico. Three oil extraction methods were compared: two traditional solvent extractions (Soxhlet and ultrasound-assisted extraction); and supercritical fluid extraction. Fatty acid quantification was done by gas chromatography with a flame ionization detector (GC-FID). Traditional Soxhlet extraction (with hexane, eight hours) produced the highest yield (5.07 ± 0.09 g oil/100 g dry weight) and the most significant amount of saturated fatty acids (48.25%). Ultrasound-assisted extraction (1 h at 40 or 60°C with hexane) produced similar fatty acid yields at either temperature but had the highest unsaturated fatty acids yields of the three methods (58.57% for 40°C, and 57.85% for 60°C). Extraction with supercritical fluid (CO_2, 10 MPa, 35°C and 30 mL CO_2/min, three h dynamic extraction) resulted in the most significant percentage of linolenic acid (10.68 ± 2.43%). The supercritical fluid and ultrasound-assisted methods produced the best quality oil, which had substantial polyunsaturated fatty acids content.

Keywords: fatty acids, avocado seed, Soxhlet, ultrasound, supercritical extraction

1. INTRODUCTION

Avocado fruit (*Persea americana* Mill.) is an essential and ancient element in the Mexican diet. Mainly marketed fresh, avocado fruit is a major agricultural crop in Mexico, which supplies approximately 50% of the international market (Secretaría de Agricultura, Ganadería, Desarrollo Rural, Pesca y Alimentación [SAGARPA], 2017). Increasing demand and a variety of uses have led to the expansion of avocado cultivation in the state of Oaxaca, Mexico. It forms part of the state's diversifying and expanding agricultural sector, which is increasing incomes and improving producers' quality of life. Avocado is perceived in the region as a kind of "green gold" and represents a significant income source for the state. In 2017 Oaxaca was ninth in avocado production in Mexico, with a total production of 5000 tons of fruit harvested from 2,500 hectares, valued at fifty million pesos (≈USD 2.64 million) (SAGARPA, 2018).

Avocado fruit pulp can be eaten fresh or used as a food ingredient (e.g., guacamole), while the seed (representing 12-16% of total fruit weight) (Olaeta, Undurraga & Espinosa, 2007) can be processed to extract its oil. Even though their potential utility, avocado seeds are often discarded. This fact may begin to change since recent research has found them to contain phytochemicals with pharmacological properties with potential uses in the food and cosmetics industries. These include bioactive compounds such as fatty acids (Dabas, Shegog, Ziegler & Lambert, 2013) and polyphenolic compounds (Kosińska et al., 2012). Biological activities have also been found in avocado seed extracts, including antihyperlipidemic (Asaolu et al., 2010), antimicrobial (Leite et al., 2009) and hypoglycemic effects (Koffi, Ernest & Dodiomon, 2009; Okonta, Okonta, Aguwa, 2007).

Fatty acids (FA) are a substantial portion of lipids, one of the three major components of biological matter (along with proteins and carbohydrates) and play multiple roles in humans and other organisms (Tvrzicka, Kremmyda, Stankova & Zak, 2011). Fatty acids, be they from the diet or via dietary supplements, are critical to maintaining healthy cell growth and reproduction (Orhan & Sener, 2002). The avocado seed is a widely available FA source. Currently treated as an agricultural by-product, avocado seeds could become a resource in their own right, if more effective extraction methods can recover more FA.

The large variety of fats and oils directly and indirectly used for human consumption are most commonly extracted using traditional techniques such as mechanical pressing, steam extraction, and organic solvents. Most fat extraction techniques involve solvents such as hexane, chloroform, petroleum ether or ethyl ether, among others. These techniques can be sophisticated since some (e.g., pressing) require more than one extraction technique to remove all the oil from raw material, or an additional post-extraction step to remove traces of the solvent (e.g., organic solvents). However, traditional techniques are being replaced by green technologies such as extraction with supercritical fluids in which the use of organic solvents potentially toxic to the environment and humans is eliminated or diminished while obtaining higher yields at shorter extraction times.

A supercritical fluid (SCF) is a substance that is above its critical pressure and temperature (Figure 1) and is produced by heating a gas above its critical temperature or compressing a liquid beyond its critical pressure (Nahar & Sarker, 2012).

Carbon dioxide is the most widely used SCF because of its critical temperature (approx. 31°C) is near that of the environment. This fact makes it suitable for use with thermolabile substances such as many natural essences and pharmaceutical products. Slight compression of CO_2 produces substantial changes in its density due to the high non-ideality exhibited by this fluid. It is non-flammable and odorless (both beneficial properties for laboratory use), relatively low cost, sufficiently pure, and widely available. Supercritical CO_2 (SC-CO_2) is generally an excellent solvent for the extraction of lipophilic solutes from a wide variety of samples (Hemming & Hawthorne, 1996). Under appropriate extraction conditions, a large amount of lipidic material can be dissolved in SC-CO_2.

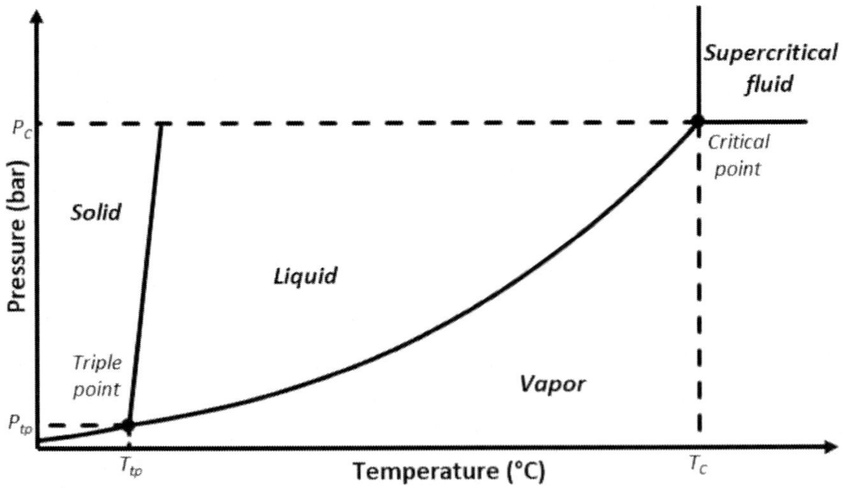

Figure 1. Phase diagram of a pure substance: P_c is critical pressure; T_c is critical temperature; P_{tp} is triple point pressure, and T_{tp} is triple point temperature.

As part of an effort to diversify the uses of avocado fruit, this chapter compares the results of different methods for extracting FA from Hass variety avocado seeds (*Persea americana* Mill.).

2. Materials and Methods

2.1. Sample Treatment

Ripe Hass variety avocado fruits were donated by an avocado production company (Oaxacan Avodeli S. P. R. de RL) in the state of Oaxaca, Mexico. These were processed by removing the seed from the pulp and exocarp. Seeds were thoroughly cleaned, grated to reduce sample size, and dried at 60 ± 2°C to constant weight in a drying oven (Felisa). The grated, dried seeds were ground in a cyclonic mill (Cyclotec™ 1093) until particle size was ≤500 micrometers (μm), and the resulting flour stored at 4°C in plastic containers until use.

2.2. Proximate Analysis

Standard techniques were used to quantify wet basis values for ash (923.03), fat (920.85) and protein (960.52) contents (Association Official Analytical Chemists [AOAC], 1995), as well as crude fiber (962.09e) and moisture (925.09) contents (AOAC, 1997). Three replicates were done of all quantifications. Nitrogen-free extract (NFE) was calculated according to FAO specifications.

Moisture content was measured by placing 5 g shredded samples in porcelain capsules at a constant weight, heating at 60 ± 2°C in a drying oven (Felisa) to constant weight, and calculating moisture content with Equation 1.

$$\text{Moisture (\%)} = \frac{W_1}{W_2} \times 100 \qquad (1)$$

where W_1 is mass loss (g), and W_2 is sample mass (g).

Ash was measured by placing 3 g sample in a crucible at a constant weight, and slowly burning it on an electric grill until combustion was complete. It was then calcined in a muffle furnace (FELISA) at 550 ± 10°C until the ash was white (24 hours). The crucible was removed from the muffle and placed in a hermetic desiccator to cool and then weighed. Ash percentage was calculated with Equation 2.

$$\text{Ash (\%)} = \frac{(W_3 - W_2)}{W_1} \times 100 \tag{2}$$

Where W_3 is mass of the crucible with ash (g), W_2 is empty crucible mass (g), and W_1 is sample mass (g).

Protein content was determined by digestion of 0.1 g sample powder with 2.5 mL concentrated H_2SO_4, in the presence of catalysts (40 mg $CuSO_4{\bullet}5H_2O$, 1.3 g K_2SO_4). Once digested, the sample was distilled in the presence of 50% NaOH (10 mL), and the distillate received in 5 mL of a solution containing 4% H_3BO_3 and bromocresol green dye. The distillate mixture was titrated with 0.1 N HCl. Calculations for total nitrogen (%) and crude protein (%) were made according to equations 3 and 4.

$$\text{Total nitrogen (\%)} = \frac{(V_2 - V_1) \times N}{W} \times 14 \times 100 \tag{3}$$

Where V_2 is the volume of HCl solution required for the problem sample (mL), V_1 is the volume of HCl solution required for the blank sample (mL), W is problem sample mass (g), and N is HCl normality in moles/1000 mL.

$$\text{Crude protein (\%)} = \text{Total nitrogen (\%)} \times \text{Conversion factor for N} \tag{4}$$

The conversion factor was 6.25.

Fat content was quantified by the Soxhlet extraction method using 2 g previously dehydrated sample in 150 mL hexane for four hours. When extraction terminated, the hexane remaining in the flask was evaporated, and the flask dried in a drying oven at 60 ± 2°C for one hour. The calculation of fat content (%) was done with Equation 5.

$$\text{Extractable fat (\%)} = [(W_3 - W_2)/W_1] \times 100 \quad (5)$$

Where W_1 is sample mass (g), W_2 is empty flask mass (g), and W_3 is flask mass with fat (g).

Crude fiber content was determined by digesting 1 g sample powder in 200 mL 1.25% H_2SO_4 for 30 minutes. The solution was filtered, the solids retained in the paper filter removed with 200 mL 1.25% sodium hydroxide, and this solution heated to boiling for 30 minutes. It was then filtered, the solids retained in the filter paper washed with boiling water and 1% HCl until the filtrate reached neutral pH, and the solids washed with alcohol and acetone. The filter paper containing the solids was recovered, placed in a porcelain crucible and dried at 100°C to constant weight, and finally incinerated at 550°C for one hour. Fiber content was calculated with Equation 6.

$$\text{Crude fiber content (\%)} = [(W_2 - W_3)/W_1] \times 100 \quad (6)$$

Where W_1 is sample mass (g), W_2 is mass of the crucible containing dried residue, and W_3 is mass of the crucible containing the ash (g).

Total carbohydrates were calculated as the difference (%) of the other components, that is, moisture (M), crude protein (CP), fats (F), ash (A), and crude fiber (CF). This value mainly represents digestible carbohydrates, in addition to vitamins and other non-nitrogen soluble organic compounds. It was expressed as a nitrogen-free extract (NFE) and calculated using Equation 7.

$$\%\text{NFE} = 100 - (\%\text{M} + \%\text{CP} + \%\text{F} + \%\text{A} + \%\text{CF}) \quad (7)$$

2.3. Oil Extraction via Soxhlet, Ultrasound and Supercritical Fluid

In the Soxhlet extraction protocol, 2.0 g dried avocado seed were placed in 150 mL hexane containing ten ppm 2.6-bis(1,7-dimethyl ethyl)-4-methyl phenol (BHT) and boiled for eight hours. For UAE, a 5.0 g sample was placed in hexane with ten ppm BHT at a 1:10 (solid: liquid, w/v) ratio for one h at 40°C or 60°C. The SFE technique was implemented using SC-CO_2 in a laboratory device built by Martínez-Hernández (2016). Briefly, a 30.0 g sample was placed in a 750 mL extraction cell and processed at 10 MPa, 35°C, and a 30 mL/min flow rate. The extraction sequence began with a 30 min equilibrium stage followed by three one-hour-long dynamic stages, each interspersed with a 15 min static stage. Three replicates were done of all three extraction techniques.

2.4. Identification and Quantification of Avocado Seed Fatty Acid

Composition of the Hass variety avocado seed FA was determined by gas chromatography with a flame ionization detector (CG-FID) using a Perkin Elmer Auto system chromatographer, and an Agilent Zebron-ZB-Wax column (30 m x 0.25 mm, 0.25 μm). Fatty acid standards for stearic, palmitic, myristic, myristoleic, palmitoleic, oleic, linoleic, and linolenic acids were acquired from Sigma Aldrich. Analytical-grade (HPLC) hexane, KOH, and HCl were acquired from Burdick Jackson, Merck.

The extracts and standards were derivatized as methyl-esters according to the following procedure. Hexane (1 mL) and 10 mL MeOH with 5% HCl were added to 200 μL extract. The mixture was heated at 70°C for two h and cooled to room temperature. Water (5 mL) and 2 mL hexane was added, the mixture transferred to a separating funnel and shaken vigorously. The bottom water was discarded, and the hexane phase completely evaporated under a nitrogen stream. This dried phase was diluted to 1 mL (SFE samples were completed to 2 mL), and 2 μL injected into the chromatographer following established conditions. Nitrogen was the carrier gas and was used a nine psi flow rate. Detector port temperature was 250°C and injector port temperature 230°C. The column temperature was programmed to ramp up from 120°C (5 min) to 220°C (5°C/min) and finally to 250°C (2°C/min, 2 min). Fatty acid methyl esters (FAMEs) in the extracts were identified by comparing retention times with the FAMEs of the standards. Quantification was done with calibration curves prepared for the identified FA.

2.5. Statistical Analysis

Both the values of the extraction yields obtained with the extraction methods and the quantities of identified FA were compared using the Student's t-statistic with a $p \leq 0.05$ value.

3. Results and Discussion

3.1. Avocado Seed Proximate Composition

The analyzed Hass variety of avocado seeds were primarily composed of water and digestible carbohydrates (Table 1). The seeds contained almost 5% protein, with lesser proportions of minerals, fats, and crude fiber.

Fat content (1.73%) is comparable to that reported elsewhere, such as the 1.1 ± 0.15% (corresponding to 63.6%) recorded in Hass variety avocado seeds from the United States (Takenaga, Matsuyama, Abe, Tori & Itho, 2008). In contrast, the present fat content constitutes 78.6% of the 2.20 ± 0.0% reported for Hass variety avocado seeds from Guatemala (Universidad del Valle de Guatemala [UVG], 2009). In this study growing conditions probably accounted for other differences in seed component proportions: 60.14 ± 0.31% moisture; 1.53 ± 0.01% ash; 1.37 ± 0.01% crude protein; 1.58 ± 0.18% crude fiber; and 31.70 ± 0.22% NFE. The present fat content results agree with the approximately 2% oil content generally reported for avocado seeds (Qin & Zhong, 2016).

Table 1. Avocado seed proximate chemical composition

Nutrient	Composition (g/100 g)
Moisture	49.03 ± 0.41
Ash	2.60 ± 0.01
Fat	1.73 ± 0.21
Crude protein	4.50 ± 0.50
Crude fiber	0.92 ± 0.10
NFE	41.36 ± 0.27

3.2. Extraction Yields

The yield was highest with the Soxhlet technique, followed by the UAE and SFE (Table 2). These differences may occur because the extraction time used in the Soxhlet method was longer than in the other extraction methods. If the Soxhlet yield is treated as 100%, the SFE yield corresponds to 9.00 ± 0.74 g oil/100 g d.w. Although the SFE technique produces lower yields than the two conventional techniques used here, the resulting extracts are of higher purity and quality due to the low temperatures used in the extraction.

Higher yields have been reported for avocado seed oil extraction by Soxhlet (7.64 ± 0.17%) and SFE with SC-CO_2 (5.33 ± 0.09%) (García-Fajardo, Ramos-Godínez & Mora-Galindo, 1999). The differences can be attributed to fruit origin and extraction conditions; for instance, pressure in this study was 27.58 MPa (2.8% higher than in the present study), and the temperature was 60°C.

Table 2. Extraction yield of Hass variety avocado seed oil from three different techniques

Extraction method	Solvent	Extraction conditions	Yield (g oil/100 g, d.w.)
Soxhlet	C_6H_{14}	8 h	5.07 ± 0.09
UAE	C_6H_{14}	1 h, 40°C	2.01 ± 0.05
		1 h, 60°C	1.90 ± 0.12
SFE	CO_2	3 h dynamic stage, 35°C, 100 bar	0.46 ± 0.04

d.w. = dry weight; UAE = ultrasound-assisted extraction; SFE = supercritical fluid extraction.

3.3. Identification and Quantification of Avocado Seed Fatty Acid

The most considerable amount of saturated fatty acids (SFA) (48.25%) was extracted with the Soxhlet technique, and the principal components in this extract were oleic acid (42.68 ± 4.78%) and palmitic acid (40.50 ± 9.62%) (Table 3). This composition differs from that reported for Hass variety avocadoes cultivated in Peru, which had high linoleic acid (48.77%) and linolenic acid (12.17%) contents (Flores et al., 2019). Unsaturated fatty acids yield was highest in both UAE treatments (58.57% at 40°C, and 57.85% at 60°C). The SFE produced the most significant percentage of linolenic acid (10.68 ± 2.43%). Polyunsaturated oil content, an indicator of the oil quality, was highest with the UAE (38.00%) and the SFE (37.00%). This fact agrees with an evaluation of the quality of avocado oil extracted by Soxhlet and SC-CO_2 in which the supercritical fluids technique produced the highest quality oil (Satriana, Supardan, Arpi & Wan Mustapha, 2019; Restrepo et al., 2012).

In a previous study, avocado seed FA was found to contain 38.9% linoleic acid and 6.57% linolenic acid (UVG, 2009). The former is higher than the highest value observed in the present results (31.43 ± 3.06% in UAE at 60°C), but linolenic acid contents were higher in both UAE techniques and the SFE in the present study (linolenic acid was not detected with the Soxhlet method). These levels contrast with a previous study in which hexane extraction produced a 20.8% concentration containing 27 FA in avocado seed oil (Fuerte variety), and palmitic acid was the main SFA (Bora, Narain, Rocha, & Queiroz, 2001).

Avocado oil quality depends on fruit quality and maturity, as well as extraction techniques in terms of temperature, solvents used, and conservation methods (Flores et al., 2019). Differences in FA composition between different extraction techniques can be attributed to the temperature and solvent used.

Table 3. Fatty acids composition of Hass variety avocado seeds by different extraction techniques

	Fatty acids composition (%)			
Name/Method	**Soxhlet**	**UAE 40°C**	**UAE 60°C**	**SFE**
Myristic, C14:0	ND	0.60 ± 0.20	0.38 ± 0.10	0.75 ± 0.07
Palmitic, C16:0	40.50 ± 9.62	11.25 ± 0.66	11.47 ± 1.34	33.63 ± 1.05
Stearic, C18:0	7.75 ± 1.30	1.75 ± 0.27	2.13 ± 0.34	9.20 ± 0.36
Oleic, C18:1	42.68 ± 4.78	20.57 ± 2.64	19.80 ± 2.78	12.38 ± 0.71
Linoleic, C18:2	4.88 ± 1.88	30.78 ± 2.68	31.43 ± 3.06	26.33 ± 2.76
Linolenic, C18:3	ND	7.22 ± 0.56	6.62 ± 0.39	10.68 ± 2.43
Saturated	48.25	13.60	13.98	43.58
Unsaturated	47.56	58.57	57.85	49.38
Polyunsaturated	4.88	38.00	38.05	37.00

ND = Not detected.

The essential fatty acids cannot be synthesized by the human organism and must be acquired from the diet. Avocado seed oil obtained by SFE using SC-CO_2 is an excellent source of essential fatty acids, suggesting it could be of interest to the food industry. The SFE extracts are particularly promising because they can be added directly to final products (e.g., food) without further treatment. After all, the solvent is completely removed from the extract during decompression. The SFE also resulted in relatively shorter extraction times and did not require the use of organic solvents. Indeed, SFE is preferred to hexane extraction, and the SC-CO_2 technique is reported to be better than hexane as a solvent for the extraction of avocado oil (Botha & McCrindle, 2004).

Avocado seeds contain components of interest, such as fatty acids. These are rich in oleic acid (a monounsaturated fatty acid), which makes them oxidation stable (Gidigbi, Ngoshe, & Martins, 2019) and, therefore, apt for industrial processing. The fatty acid profile of avocado seed oil also has a higher concentration of polyunsaturated fatty acids than pup oil (Bora et al., 2001).

CONCLUSION

Fatty acids composition in the oil from Hass variety avocado seeds was influenced by the extraction method, probably due to the differential solubilities of various fatty acids in hexane and SC-CO_2. Soxhlet extraction was used as a reference method to compare the yield of the other solid-liquid extraction methods. The non-conventional UAE and SFE extraction techniques are more environmentally friendly due to their limited use of organic solvents. Extraction with supercritical fluid did not provide the highest yields, although the overall quality of the resulting fatty acids was superior to the other tested methods. Differences in the SC-CO_2 solubility of certain desired products and other lipid-soluble bioactive compounds can be controlled by manipulating extraction conditions such as pressure and temperature. Supercritical fluid extraction can also reduce separation process cost and time compared to conventional solvent extraction when purifying specific components of interest from an extract. Future research needs to concentrate on increasing fatty acids solubility in the SC-CO_2 technique by applying higher pressure and temperature conditions during extraction. Avocado seeds are a plentiful raw material most often treated as agricultural waste, but the present results clearly show them to be an excellent source of unsaturated fatty acids. Several studies have addressed the application of SFE in obtaining avocado pulp oil, but the present study is the first to identify the different fatty acids in avocado seed extracts obtained with the SC-CO_2 technique.

REFERENCES

Association Oficial Analytical Chemists (AOAC). 1995. *Official Methods of Analysis*. Washington, D. C., USA.

Association Oficial Analytical Chemists (AOAC). 1997. *Official Methods of Analysis*. Washington, D. C., USA.

Asaolu, M. F., Asaolu, S. S., Fakunle, J. B., Emman-Okon, B. O., Ajayi, E. O. & Togun, R. A., 2010. "Evaluation of in-vitro antioxidant activities of methanol extracts of *Persea americana* and *Cnidosculus aconitifolius*." *Pak. J. Nutr.*, *9*(11), 1074–1077. https://dx.doi.org/10.3923/pjn.2010.1074.1077.

Bora, P., Narain, N., Rocha, R., & Queiroz, P. M., 2001. "Characterization of the oils from the pulp and seeds of avocado (Fuerte cultivate) fruits." *Fats and Oils, 52*(3-4), 171-174. https://dx.doi.org/10.3989/gya.2001.v52.i3-4.353.

Botha, B.M. & McCrindle, RI, 2004. Supercritical fluid extraction of avocado oil. *South African Avocado Growers' Association Yearbook*, 27, 24-27.

Dabas, D., Shegog, R. M., Ziegler, G. R., & Lambert, J. D., 2013. "Avocado (*Persea americana*) seed as a source of bioactive phytochemicals." *Curr. Pharm. Design.*, *19*, 6133-6140. https://dx.doi.org/10.2174/1381612811319340007.

Flores, M., Saravia, C., Vergara, C., Avila, F., Valdés, H., & Ortiz-Viedma, J., 2019. "Avocado oil: Characteristics, properties, and applications." *Molecules*, *24*(11), 2172. http:s//dx.doi.org/10.3390/molecules24112172.

García-Fajardo, J. A., Ramos-Godínez, M. del R., and Mora-Galindo, J. 1999. "Structure of the avocado seed and quantification of the fat extracted by different techniques." *Rev. Chapingo Ser. Hortic., 5*, 123-128. Retrieved from https://www.chapingo.mx/revistas/revistas/articulos/doc/rchshV1104.pdf.

Gidigbi, J. A., Ngoshe, A. M. & Martins, A. 2019. "Industrial viability study of the avocado seed oil." *International Journal of Recent*

Innovations in Academic Research, 3(6), 48-57. Retrieved from https://www.ijriar.com/docs/volume3/issue6/IJRIAR-05.pdf.

Hemming, F. W., and Hawthorne, J. N., 1996. *Lipid Analysis*. London, United Kingdom: Taylor & Francis Ltd.

Koffi, N., Ernest, A. K. & Dodiomon, S. 2009. "Effect of aqueous extract of *Persea americana* seeds on the glycemia of diabetic rabbits." *Eur. J. Sci. Res.*, *26*(3), 376–385.

Kosińska, A., Karamać, M., Estrella, I., Hernández, T., Bartolomé, B., & Dykes, G. 2012. "A phenolic compound profiles and antioxidant capacity of *Persea americana* Mill. Peels and seeds of two varieties." *J. Agric. Food Chem., 60*, 4613–4619. https://dx.doi.org/10.1021/jf300090p.

Leite, J. J. G., Brito, E. H. S., Cordeiro, R. A., Brilhante, R. S. N., Sidrim, J. J. C., Bertini, L. M., de Morais & S. M., Rocha, M. F. G. 2009. « Chemical composition, toxicity and larvicidal and antifungal activities of *Persea americana* (avocado) seed extracts." *Rev. Soc. Bras. Med. Trop.*, *42*(2), 110–113.

Martínez-Hernández, U. P., 2016. *Design, construction, and experimental validation of a laboratory scale supercritical extraction equipment*. (Undergraduate thesis). Technological University of the Mixteca, Oaxaca, Mexico.

Nahar, L., & Sarker, S. D., 2012. "Supercritical fluid extraction in natural product analyses." *Methods Mol. Biol., 864*, 43-74. https://dx.doi.org/10.1007/978-1-61779-624-1_3.

Okonta, M., Okonta, L. & Aguwa, C. N., 2007. "Blood glucose low- ering activities of seed of *Persea americana* on alloxan induced diabetic rats." *Nig. J. Nat. Prod. Med.*, *11*, 26–28. https://dx.doi.org/10.4314/njnpm.v11i1.11874.

Olaeta, J. A., Undurraga, P., & Espinosa, G. November 2007. "Evolution of the content of oil and unsaponifiable compounds in Hass, Fuerte and Isabel avocados (*Persea Americana* Mill.)." *Proceedings VI World Avocado Congress*. Viña del Mar, Chile.

Orhan, I., & Sener, B., 2002. "Fatty acid content of selected seed oils." *J. Herb. Pharmacother.*, *2*(3), 29-33. https://dx.doi.org/10.1080/j157v02n03_03.

Qin, X., & Zhong, J., 2016. "A review of extraction techniques for avocado oil." *J. Oleo Sci.*, *65*(11), 881–888. https://dx.doi.org/10.5650/jos.ess16063.

Restrepo, D. A. M., Londoño, L. J., González, Á. D., Benavides, P. Y. & Cardona, S. B. L. 2012. "Comparison of Hass variety avocado oil grown in Colombia, obtained by supercritical fluids and conventional methods: a quality perspective." *Rev. Lasallista Investig.*, *9*(2), 151-161. Retrieved from https://www.redalyc.org/articulo.oa?id=695/69525875005.

Satriana, S., Supardan, M. D., Arpi, N., & Wan Mustapha, W. A., 2018. Development of methods used in the extraction of avocado oil. *Eur. J. Lipid, Sci. Tech.*, *121*, 1800210. https://dx.doi.org/10.1002/ejlt.201800210.

Secretaría de Agricultura, Ganadería, Desarro Rural, Pesca y Alimentación (SAGARPA). 2017. *National Agricultural Planning 2017-2018, Mexican Avocado.* Retrieved from https://www.gob.mx/cms/uploads/attachment/file/257067/Potencial-Aguacate.pdf.

Secretaría de Agricultura, Ganadería, Desarrollo Rural, Pesca y Alimentación (SAGARPA). 2018. *Monthly newsletter of the avocado production.* Retrieved from https://www.gob.mx/cms/uploads/attachment/file/302238/Boletin_mensual_producci_n_de_aguacate_enero_2018.pdf.

Takenaga, F. Matsuyama, K. Abe, S. Torii, Y. & Itoh, S. 2008. "Lipid and fatty acid composition of mesocarp and seed of avocado fruits harvested at northern range in Japan." *J. Oleo Sci.*, *57*(11), 591-597. https://dx.doi.org/10.5650/jos.57.591.

Tvrzicka, E., Kremmyda, L. S., Stankova, B. & Zak, A. 2011. "Fatty acids as biocompounds: their role in human metabolism, health, and disease – a review. Part 1: classification, dietary sources and biological functions." *Biomed. Pap. Med. Fac. Univ. Palacky Olomouc Czech Repub.*, *155*(2), 117-130. https://dx.doi.org/10.5507/bp.2011.038117.

Universidad del Valle de Guatemala (UVG). 2009. *The chemical composition, antioxidant capacity and nutritional value of avocado varieties seed*. Retrieved from http://glifos.concyt.gob.gt/digital/fodecyt/fodecyt%202006.02.pdf.

In: Avocado: Consumption and Health ISBN: 978-1-53618-204-0
Editors: M. G. Melgar Lalanne et al. © 2020 Nova Science Publishers, Inc.

Chapter 11

CHIPOTLE CHILI OLEORESIN EXTRACTED WITH AVOCADO OIL AS ADDITIVE IN AMARANTH BARS

Karla Quiroz Diego[1], Guiomar Melgar Lalanne[2], Frixia Galán Méndez[1,*] and Cid Ramón González-González[3,†]

[1]Facultad de Ciencias Químicas, Universidad Veracruzana, Xalapa, Ver., México
[2]Instituto de Ciencias Básicas, Universidad Veracruzana, Xalapa, Ver., México
[3]Instituto Tecnológico Superior de Acayucan, Acayucan Ver., Mexico

[*] Corresponding Author's Email: fgalan@uv.mx.
[†] Corresponding Author's Email: cidgonzalez@itsacayucan.edu.mx.

ABSTRACT

Chipotle chili oleoresin (CCO) extracted with avocado oil has shown antimicrobial activity and oxidative stability. Furthermore, amaranth is a native Mexican crop, and it has a growing interest by consumers due to its high-quality protein content. The objective of the present study was to formulate and evaluate an amaranth bar enriched with Chipotle chili oleoresin (CCO) extracted in avocado oil as an additive. Three different formulations were tested: A) control bar without CCO, B) added with 7 mL/100 g, and C) 14 mL/100 g of CCO. Microbial analysis was carried out to identify aerobic mesophilic microorganisms, total coliforms, and yeast and mold according to the methods described in the Mexican Official Standards (NOM).

Additionally, a sensory evaluation was carried out with non-trained panelists using a 9-point Hedonic scale. Finally, we carried out a proximate analysis of the amaranth bar C as well as its shelf-life stability, comparing physicochemical parameters such as pH, titratable acidity, moisture, and peroxide value. These parameters were tested at the 3rd and 15th days of storage at both 45 and 28°C. The results showed that the addition of CCO improved the acceptability of the product and provided technological advantages, such as enhancing shelf-life inhibiting microbial growth and showing oxidative stability. This formulation takes advantage of the CCO using avocado oil as an extracting solvent and the nutritional value of amaranth and cranberry, generating a gluten-free product rich in proteins with an enhanced shelf-life.

Keywords: Chipotle chili oleoresin, avocado oil-extraction, amaranth bar

1. INTRODUCTION

The cold-pressed avocado (*Persea americana Mill.*) oil is a high-quality oil and free from organic solvent traces (Aydeniz Güneser, Yilmaz, & Ok, 2017). It possesses oxidative stability, and no addition of synthetic antioxidants are needed (Cerecedo-Cruz, Azuara-Nieto, Hernández-Álvarez, González-González, & Melgar-Lalanne, 2018). Besides, the oleoresin fraction of Chipotle chili pepper (smoked jalapeno; *Capsicum annuum* L.) is reported to contain capsaicin and dihydrocapsaicin, 883 mg/kg and 552 mg/kg, respectively (Orellana-Escobedo, Garcia-Amezquita, Olivas, Ornelas-Paz, & Sepulveda, 2013), which are the compounds responsible for the 90% of the pungency (Iwai, Suzuki, Fujiwake, & Iwai, 1979). The use of oleoresins in the food industry provides certain advantages, such as the improvement of product stability due to its antimicrobial and antioxidant properties (Omolo, 2014; Restrepo-Gallego, 2007). Moreover, the use of cold-pressed avocado oil as a means for extracting oleoresins provides the technological advantage of achieving a high yield of bioactive compounds by cold extraction (25°C) (Cerecedo-Cruz et al., 2018). Thus, Chipotle chili oleoresin (CCO) extracted with avocado oil may be used as an additive for enhancing the shelf-life of food products.

The amaranth, *Amaranthus ssp* (A.), is historically a highly appreciated crop in The Americas (Rastogi & Shukla, 2013; Soriano-Santos & Avendaño-Martinez, 2013). This genus comprehends three species: *A. hypochondriacus, A. cruentus* (grown in Mexico), and *A. caudatus* (cropped in Peru) (Bressani, 2006). The protein content is 14-18% of dry weight, which is higher than that of traditional cereals such as maize, rice, and wheat (Soriano-Santos & Avendaño-Martinez, 2013). Nonetheless, its nutritional value is rooted in its protein quality, not its protein quantity. Amaranth contains seven of nine essential amino acids: Leu, Lys, Val, Met, Phe, Thr, and Ile (Barba de la Rosa et al., 2009). The amounts of Hys, Leu, Thr, Trp, and sulfur and aromatic amino acids found in amaranth varieties cover the daily dietary requirements recommended by WHO/FAO/UNU (Aguilar et al., 2013; WHO/FAO/UNU, 2007). Furthermore, it contains more lipids and minerals than traditional cereals, and it is an essential source of squalene, ω3 (alpha-linoleic acid), and ω6 (linoleic acid) (Aguilar et al., 2013). The starch represents about 60-66% of its dried mass (Srichuwong et al., 2017). Indeed, amaranth is considered by the FAO as one of the pseudocereals with great potential to fight the food insecurity challenge worldwide (FAO, 2013). Despite all these nutritional advantages, the amaranth is not widely consumed due to the lack of promotion and technological developments that could facilitate its consumption (Ayala-Garay, Espitia-Rangel, Rivas-Valencia, Martínez-Trejo, & Almaguer-Vargas, 2016; Soriano-Santos & Avendaño-Martinez, 2013). For instance, the lack of gluten disqualifies its individual use in the baking process, although it can be used to enrich flours enhancing their nutritional value (Singh, Singh, Shevkani, & Virdi, 2019). In México, amaranth is traditionally consumed as a candy bar called "alegría," which is a food product enriched with saccharose; however, this may represent a nutritional drawback due to its high glycemic index. Therefore, more research and technological development are needed to take advantage of amaranth's nutritional value.

The cranberry (*Vaccinium macrocarpon*) is a small fruit with a sour and bitter taste. It contains bioactive compounds that provide health benefits to their consumers, such as antioxidant and anticarcinogenic polyphenols (He & Rui, 2006; Kalın, Gülçin, & Gören, 2015; Neto, 2007). Protein bars are food products that have gained importance in the food market recently, and it is a product where amaranth could be easily integrated as a food ingredient in a similar way to the "alegría" snack. Thus, the objective of this study was to formulate and characterize the physicochemical and sensory properties of an amaranth bar mixed with a cranberries paste as a binding agent and CCO extracted with avocado oil as an additive. We were aiming for a nutritionally balanced food product with low simple carbohydrates and with an enhanced shelf-life.

2. Material and Methods

2.1. Oleoresin Extraction

To extract the oleoresin from Chipotle chili pepper, we followed the methodology as described by Cerecedo et al., (2018), shown in Figure 1. Firstly, the chili peppers were washed and then dried in an oven (Humboldt MFG, 40 Lab Oven, USA) at 45°C for 48 h. The dried fruit was ground in a conventional miller and was then centrifuged (Hermle, Z 300 K, Germany) three times at 3250 rpm for 15 min, reaching a concentration of 100 g of chili powder/ 200 g of cold-pressed avocado oil (bought in the supermarket as "virgin oil") (Ahuacatlán, Mexico). The supernatant was then separated and stored in an amber flask under 10°C until further experiments.

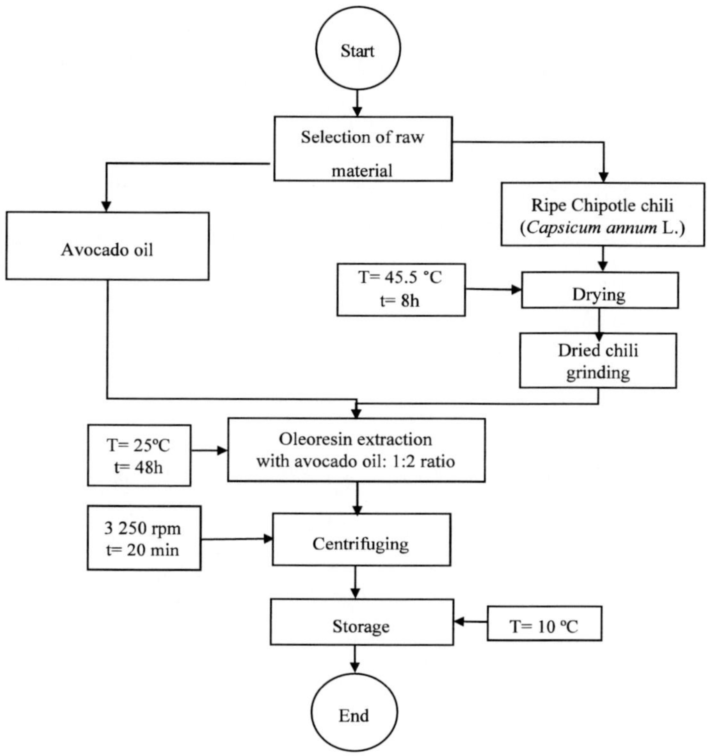

Figure 1. Oleoresin from Chipotle chili extraction with avocado oil (Modified from Cerecedo, 2016).

2.2. Amaranth Bar Production

The methodology to process the bar described by Velastegui (2016) was followed with some modifications, as shown in Figure 2. The first step was to prepare the binding agent, which consisted of a paste made of dried cranberries. Cranberries were chosen for their sensory properties and the ability to bind the amaranth seeds with a firm consistency, molding the bars with ease. The cranberries were soaked with filtered water and then blended in a food processor until obtaining a homogeneous mixture. It was then mixed manually with the CCO in a stainless-steel container. The first formulation, labeled as A, did not have CCO added. Amaranth bars B and C had 7 mL and 14 mL of CCO respectively per 100 g of bar mix. 200 g of amaranth seeds were added to 200 g of cranberry paste and CCO formulations and then molded into small bars of 5 * 3 cm with the aid of a stainless steel cutter. The bars were baked in a pre-heated conventional oven at 232°C for 5 min on a tray covered with wax paper. The bars cooled down at room temperature (23 ± 2°C) for 20 min and then were packed into polypropylene sealing bags.

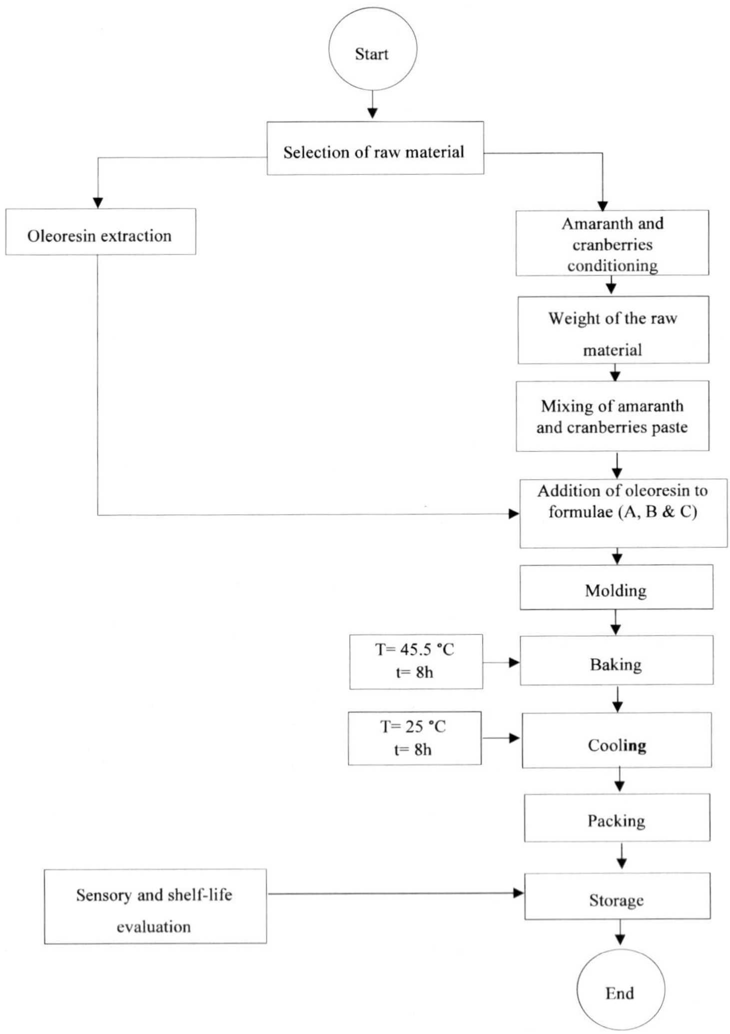

Figure 2. Process of formulating an amaranth bar.

2.3. Microbial and Sensory Analysis

The microbial analysis was carried out according to the Mexican Official Standards (NOM) and consisted of mesophilic aerobic count ((NOM-092-SSA1-1994, 1994), total coliform count (NOM-113-SSA1-1994, 1994) and yeast and mold count (NOM-111-SSA1-1994, 1994). Firstly, 2 g of each bar were sampled, and 10-folded dilutions were plated out on their respective agar (BD, Biocon, Mexico) according to the NOM and incubated at 37°C for 24 hrs for colony counting.

A 9-point Hedonic scale, 1: dislike extremely, 2: dislike very much, 3: dislike moderately, 4: dislike slightly, 5: acceptable, 6: like slightly, 7: like moderately, 8: like very much, 9: like extremely, was used to perform the test for acceptability. Fifty-five non-trained panelists between 17 and 48 years old took part in the study, from which 60% were female and 40% male. In front of each panelist, a 5 g sample of each bar type was presented (A, B, and C). Proximate analysis was performed to the bar with higher acceptance (C), and its shelf-life was estimated compared to control (A).

Proximate analysis was carried out on bar C according to the following standards: total fat by the Soxhlet extraction method (NMX-F-615-NORMEX-2004, 2004), crude fiber (AOAC, 2006), total carbohydrates (López-Legarda, Taramuel-Gallardo, Arboleda-Echavarría, Segura-Sánchez, & Restrepo-Betancur, 2017), reducing sugars by de 3,5-dinitrosalicylic acid method (Lindsay, 1973), crude protein by micro Kjeldahl (NOM-F-68-S-1980, 1980), moisture (NMX-F-083-1986, 1986) and ashes (NMX-F-066-S-1978, 1978); all the essays were made by triplicate.

Shelf-life was evaluated under stressed conditions, a relative humidity of 89% for 3 and 15 days at 45°C and 28 ± 2°C, respectively. Samples for microbial analysis were taken at days 0, 3, 7, and 15, monitoring aerobic mesophilic, total coliforms, yeast and mold, pH, acidity, moisture, and peroxide value (PV).

The PV was determined according to the AOCS official methods Cd 8–53 (AOCS, 1993). Firstly, 2.5 mL of acetic acid/dichloromethane solution (3:2 v/v) was added to 0.5 g of the sample. Secondly, 0.05 mL of a saturated solution of KI in dark conditions was added and left for 1 min. Finally, 7.5 mL of cold distilled water was added. When a blue color appeared, the solution was titrated with $Na_2S_2O_3$ 0.001 N, until the blue color disappears. PV was expressed as peroxide milliequivalents of active oxygen per kilogram of bar dry mass (meqO_2/kg). PV was only determined in bar C at 28°C, in days 0 and 15. The nutritional bar with no oleoresin was used as control (A).

An analysis of variance (ANOVA) was performed, and a Tukey test among the different formulations ($\alpha = 0.05$). For the sensory evaluation, an ANOVA was performed using the index of acceptability as the dependent variable.

3. Results and Discussion

The present work consisted of comparing three formulations of amaranth bars containing, respectively, 0, 7, and 14 mL of CCO per 100 g of bulk mass consisting of 50% of amaranth seeds and 50% of dried cranberry paste. Firstly, microbiological and sensory assessments were carried out. The results showed that the microbial growth was lower in the bars with added CCO, and the response was dependant on the concentration of CCO added, as shown in Table 1. Aerobic mesophilic counts were lower in higher concentrations of CCO, and no growth of fungi or total coliforms were detected in the sample of amaranth bar C. The counts were lower than the upper limits established by the Mexican Offical Standards (NOM-247-SSA1-2008) in all cases.

The results of the sensory test by non-trained panelists showed that CCO enhanced the general acceptability of the bars. The analysis of variance in Table 2 indicated that there is a significant difference between the formulations compared to the control. Furthermore, the p-values in the Tukey post-hoc analysis showed significant differences among the attributes tested, where amaranth bar C showed the highest general acceptability.

Table 1. Microbiological analysis of amaranth bar formulations: A) control without oleoresin; added with B) 7 mL/100 g; and C) 14 mL/100 g of Chipotle chili oleoresin. Results are expressed as the mean ± SD; n = 3

Analysis	Bar formulations			
	A	B	C	Upper limits*
Aerobic mesophilic ($\log_{10}$ CFU/g)	3.498 ± 1.845	2.740 ± 1.845	2.544 ± 1.447	4.00
Yeast and mold ($\log_{10}$CFU/g)	1.176 ± 0.845	1.000 ± 0.000	ND	2.477
Total coliforms ($\log_{10}$CFU/g)	ND	ND	ND	<1.47

* As published in the Mexican Official Standard NOM-247-SSA1-2008.
ND: No detected. Detection limit was one $\log_{10}$ CFU/g.

Table 2. Analysis of variances of general acceptability of amaranth bar formulations added with Chipotle chili oleoresin: A) 0 mL/100 g, B) 7 mL/100 g and C) 14 mL/100g of the total mass

	Degrees of freedom	Sum of squares	Mean of squares	F-value	P-value
Total	11	2.9159	0.2651		
Formulation bar	2	2.5581	1.2790	94.2005	<0.0001
Subjects (relation)	3	0.2764	0.0921	6.7846	0.0235
Error	6	0.0815	0.0136		

In Figure 3, it can be observed that bar C had the highest appreciation with a score of 7 (“I like it moderately”) in the attributes of color, aroma, and texture. The flavor was the attribute with more variability, where bar A scored 5 “Acceptable”; bar B scored 7 “I like it moderately” ($P < 0.001$) and formulation C was 8 “I like it very much” ($P < 0.001$). Indeed, the control bar A had the least acceptability score compared to bar B and C. Moreover, texture perception was improved significantly in both bars compared to the control A ($P < 0.001$). The panelists commented that bar C had the best consistency. These results may be due to the effect of the content of avocado oil on the baked bar, which may improve the texture.

Table 3. Tukey test of sensory attributes of amaranth bar formulations added with Chipotle chili oleoresin extracted with avocado oil: A) 0 mL/100 g, B) 7 mL/100g and C) 14 mL/100g of total mass (n = 55)

			Difference of means	95% CI[a]			
Bar formulation - Attributes				**lower**	**upper**	**P-value**	Sig.
B aroma	-	A aroma	0.4923	-0.1553	1.1399	0.3463	
C aroma	-	A aroma	0.7077	0.0601	1.3553	0.0186	*
C aroma	-	B aroma	0.2154	-0.4322	0.8630	0.9951	
B color	-	A color	0.4154	-0.2322	1.0630	0.6209	
C color	-	A color	0.7846	0.1370	1.4322	0.0044	**
C color	-	B color	0.3692	-0.2783	1.0168	0.7781	
B flavor	-	A flavor	1.1692	0.5217	1.8168	0.0000	***
C flavor	-	A flavor	1.7692	1.1217	2.4168	0.0000	***
C flavor	-	B flavor	0.6000	-0.0476	1.2476	0.0999	
B texture	-	A texture	0.9385	0.2909	1.5860	0.0002	***
C texture	-	A texture	1.2923	0.6447	1.9399	0.0000	***
C texture	-	B texture	0.3538	-0.2937	1.0014	0.8226	

[a] CI: confidence interval.
Sig: significance codes: <0.001 '***', <0.01 '**', <0.05 '*'.

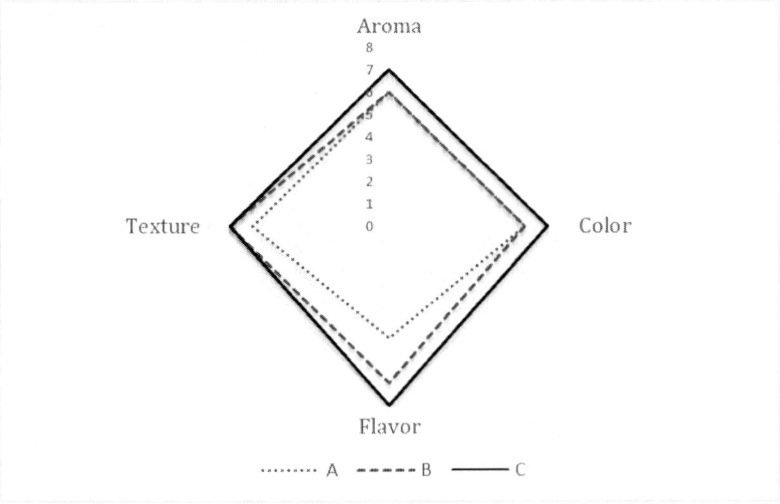

Figure 3. Sensory evaluation of different formulations of amaranth bars: A) Control 0 mL/100g, B) Oleoresin 7 mL/100 g, and C) Oleoresin 14 mL/100g.

Table 4. Proximate analysis of amaranth bar added with Chipotle chili oleoresin (14 mL/100 g). Results are the mean ± SD

Analysis	**g/100 g of dry mass**
Fat	4.947 ± 0.353
Crude fiber	9.615 ± 0.473
Total Carbohydrates	65.995 ± 2.057
Reducing sugars	8.201 ± 0.303
Crude proteins	14.901 ± 2.340
Moisture	11.564 ± 1.055
Ash	1.738 ± 0.131

Based on the results of the sensory evaluation, amaranth bar C was selected for further analysis. The proximate analysis results of bar C are shown in Table 4. The fat was determined as 4.947 ± 0.353 g/100g. These results support the inclusion of the formulated bar in a low-fat diet provided that the daily intake recommended should not exceed 30% (Garza Benito, Ferreira Montero, & del Río Ligorit, 2006). Other studies with amaranth bars have reported higher fat content such as Velastegui (2016), who reported 16.95 g/100g in a bar made of amaranth, quinoa, and moringa. Cornejo, Gaido, & López (2017) reported 10.46 g/100g in a snack made of amaranth flour and spirulina, and Velásquez & Llacho (2015) reported eight g/100g of fat in a bar made of tarwi (*Lupinus mutabilis*), kiwicha (*A. caudatus*) and wheat flour.

The crude fiber output was 9.565 ± 0.473 g/100g, which was higher than that reported by Cornejo et al., (2017) of 6.69 g/100g. This result was likely due to the fiber contained in the cranberries used as a binding agent. Cranberry pomace is reported to contain 65.5% of insoluble fiber and 5.7% of soluble fiber (White, Howard, & Prior, 2010). The consumption of fiber-rich food is essential for the prevention of diseases related to the metabolic syndrome. According to the WHO/FAO and EFSA, 25 g/d of dietary fiber intake is recommended (Jones, 2014). Thus, 100 g of a bar C formulation would cover approximately 30% of the recommended daily intake.

Total carbohydrates were determined as 65.995 ± 2.057g/100g. There are not many references reported for total carbohydrates. Nevertheless, Velastegui (2016) reported a bar with 58.87 g/100g. The composition of reducing sugars, which are simple carbohydrates of rapid assimilation, was determined as 8.201 ± 0.303 g/100g, which represents approximately 12.42% of the total carbohydrates in the bar, which means the bar has a low glycemic index.

The crude protein content was 14.901 ± 2.340 g/100 g, which is higher than that reported by Velastegui (2016) of 11.10 g/100 g with the formulation described above. The protein bars available on the market provide proteins in the range of 10 to 35 g/100g (Banach, Clark, & Lamsal, 2016). However, most of them are based on whey or soy proteins that may represent a disadvantage for those individuals that suffer lactose intolerance or may undergo food allergy caused by soy proteins or carbohydrates (Cordle, 2004). Thus, amaranth bars may represent an alternative for gluten-free protein intake and lower allergenicity (Cárdenas-Torres et al., 2019).

The moisture content was 11.564 ± 1.055 g/100g, similar to that of 12.01 g/100g reported by (Velastegui, 2016). The low moisture is directly correlated to its high stability and shelf life, provided that low moisture halts the microbial growth related to food spoilage and oxidative processes (Vera Zambrano, Dutta, Mercer, MacLean, & Touchie, 2019).

The ash content was determined as 1.738 ± 0.131 g/100 g. This value was lower than other values reported, for instance, Gómez-Flores, Gómez-Acuña, Pérez-Díaz, & Chávez-Murillo (2016) (3 g/100 g) and Cornejo et al., (2017) (6.3 g/100 g). This result could be explained by the ingredients used in other formulations such as legumins or spirulina that contain significant amounts of calcium, phosphorus, magnesium, and iron, increasing the amount of ash (Soni, Sudhakar, & Rana, 2017). It was reported by Sanz-Penella, Wronkowska, Soral-Smietana, & Haros (2013) that amaranth contributed to an increase in the minerals in bread when wheat flour was enriched with *A. cruentus*; however, also the number of phytates contained in amaranth seeds may absorb the minerals such as calcium, zinc, and iron (Sanz-Penella et al., 2013).

Table 5. Physicochemical analysis of amaranth and cranberries bar added with Chipotle chili oleoresin: A) control (0 mL/100g), and C) (14 mL/100g) at day 0 and 3 of storage at 45°*C*. The results are the means ± SD (n = 3)

Analysis	**Day 0**		**Day 3**	
	A	**C**	**A**	**C**
pH	4.316 ± 0.035^{A}	4.530 ± 0.000^{A}	4.110 ± 0.010^{A}	4.493 ± 0.015^{B}
Moisture (g/100 g)	18.476 ± 0.098^{A}	11.560 ± 1.055^{B}	22.633 ± 0.680^{A}	12.183 ± 0.040^{B}
Titratable acidity (g oleic acid/100 g)	0.113 ± 0.003^{A}	0.097 ± 0.003^{A}	0.097 ± 0.003^{A}	0.080 ± 0.003^{B}

$^{A, B}$: indicate significant difference; Tukey ($\alpha = 0.05$).

Table 6. Microbiological analysis of amaranth and cranberries bar added with Chipotle chili oleoresin: A) control (0 mL/100g), and C) (14 mL/100g) at day 0 and 3 of storage at 45°C. The results are the means ± SD (n = 3)

Analysis	**Day 0**		**Day 3**		
	A	**C**	**A**	**C**	**Upper limits***
Aerobic mesophilic ($\log_{10}$ CFU/g)	3.299 ± 0.845^{A}	3.173 ± 0.845^{B}	3.740 ± 0.778^{A}	3.525 ± 0.699^{B}	4.00
Yeast and mold ($\log_{10}$CFU/g)	ND	ND	ND	ND	2.477
Total coliforms ($\log_{10}$CFU/g)	3 ± 0.0^{A}	ND	3.301 ± 0.0^{A}	ND	<1.47

* As published in the Mexican Official Standard NOM-247-SSA1-2008.

$^{A, B}$: indicate significant difference; Tukey ($\alpha = 0.05$).

ND: no detected. The limit of detection was one $\log_{10}$ CFU/g.

Table 5 shows the results for the analysis of the comparison of the physicochemical results of bar A and C at 45°C at three days. A significant difference was found between the different treatments and the moisture change over time. The initial and final moisture of bar A was significantly higher than in bar C. This may be associated with the addition of CCO with avocado oil and their hydrophobic nature. The changes in pH and acidity were not considered significant.

The results of the microbiological analysis show that the addition of CCO in the amaranth bar C has an inhibitory effect on total coliforms below the detection limit (10 CFU/g). The addition of CCO inhibited the microbial growth due to the antimicrobial activity of capsaicinoids (Berke & Shieh, 2012; Melgar-Lalanne, Hernández-Álvarez, Jiménez-Fernández, & Azuara, 2017).

The results of the physicochemical analysis when the bars were stored at 28ºC for 15 days are shown in Table 7. The initial pH values were between 4.19 and 4.53, showing a significant change between the two formulations between 7 and 15 days of sampling. Furthermore, there was a difference in the moisture of formulations A and C, where the bar with added CCO (C) showed less moisture than the control (A). Nonetheless, not significant changes were observed between the days. Moreover, there were no significant changes in titratable acidity. The PV showed a slight increase in the 15th day for bar C. However. This value may not show any negative effect in the quality of the product, provided that rancidity is likely to be noticed when PV is equal or higher than 20 meq/kg (Ballesteros-Torres, García-Reyes, Fuentes-Tamez, Vara-Rangel, & Cabello-Treviño, 2016; Yang et al., 2013).

The results in the microbiological analysis shown in Table 8, show statistical differences for aerobic mesophilic among the three days sampled. These values are still below the limits established in the Mexican Official Standards (NOM-247-SSA1-2008). On the other hand, the total coliforms count was below the detection limit for bar C, whereas bar A showed an increase from 1.3 to 1.4 $_{\log 10}$ CFU/g, and yeast and mold were below the detection limit in all the samples. The microbial inhibition may be attributed to both the low moisture and the antimicrobial activity induced by the oleoresin. However, the differences of the CCO-added bars compared to the control bar indicate the effectiveness of the CCO extracted with avocado oil as an antimicrobial additive.

Table 7. Physicochemical analysis of amaranth and cranberries bar added with Chipotle chili oleoresin: A) control (0mL/100g), and C) (14mL/100g) at day 0, 7 and 15 of storage at 28°C. The results are the means ± SD (n = 2)

Analysis	**Day 0**		**Day 7**		**Day 15**	
	A	**C**	**A**	**C**	**A**	**C**
pH	4.316 ± 0.035^{A}	4.533 ± 0.000^{B}	4.186 ± 0.090^{A}	4.533 ± 0.005^{B}	4.120 ± 0.020^{A}	4.523 ± 0.005^{B}
Moisture (g/100 g)	18.476 ± 0.098^{A}	11.560 ± 1.055^{B}	19.413 ± 0.246^{A}	11.820 ± 0.797^{B}	20.093 ± 0.490^{A}	11.900 ± 0.026^{B}
Titratable acidity (g oleic acid/100 g)	0.113 ± 0.003^{A}	0.097 ± 0.003^{A}	0.108 ± 0.003^{A}	0.093 ± 0.003^{A}	0.098 ± 0.003^{A}	0.091 ± 0.003^{A}
Peroxide value (meq/ kg peroxide/mass)	2.000 ± 0.000^{A}	2.150 ± 0.070^{A}	-	-	2.150 ± 0.070^{A}	2.95 ± 0.700^{B}

$^{A, B}$: indicate significant difference; Tukey ($\alpha = 0.05$).

(-) indicates that the sample was not analyzed.

Table 8. Microbiological analysis of amaranth and cranberries bar added with Chipotle chili oleoresin: A) control (0 mL/100g), and C) (14 mL/100g) at day 0, 7, *and 15 of storage at 28*°C. The results are the means ± SD (n = 2)

Analysis	**Day 0**		**Day 7**		**Day 15**		
	A	**C**	**A**	**C**	**A**	**C**	**Upper limits***
Aerobic mesophilic (log_{10} CFU/g)	3.299 ± 0.845 ^{A}	3.173 ± 0.845 ^{B}	3.322 ± 0.845^{A}	3.176 ± 0.845^{B}	3.415 ± 0.761^{A}	3.176 ± 0.778^{B}	4.00
Yeast and mold (log_{10}CFU/g)	ND	ND	ND	ND	ND	ND	2.477
Total coliforms (log_{10}CFU/g)	1.0 ± 0.903^{A}	ND	1.301 ± 0.845^{A}	ND	1.398 ± 0.845^{A}	ND	<1.47

* As published in the Mexican Official Standard NOM-247-SSA1-2008.

$^{A, B}$: indicate significant difference; Tukey ($\alpha = 0.05$).

ND: no detected. The detection limit was one log_{10} CFU/g.

Conclusion

In this study, it has been demonstrated that CCO, extracted with cold-pressed avocado oil, when used as an additive in amaranth bars formulated with cranberries, improved the final product. It increased the general acceptability of the sensory attributes and improved the microbial quality to below the limits established by the Mexican Official Standards. Additionally, the bar with CCO showed low oxidative activity after 15 days of storage at 28°C, indicating high stability. The results obtained here indicate that the avocado oil is suitable for the extraction of bioactive compounds found in the oleoresin preserving their functionality. Finally, this technological approach takes advantage of the high-quality and gluten-free protein contained in the amaranth seeds, which makes it suitable for vegans or those individuals with celiac conditions.

References

Aguilar, E. G., Peiretti, E. G., Uñates, M. A., Marchevsky, E. J., Escudero, N. L., & Camiña, J. M. (2013). Amaranth seed varieties. A chemometric approach. *Journal of Food Measurement and Characterization*. https://doi.org/10.1007/s11694-013-9156-1.

AOAC. *Official Methods of Analysis, Proximate Analysis and Calculations Crude Fiber; Procedure Ba 6a-05* (2006). The US.

AOCS. (1993). Peroxide value acetic acid-chloroform method–AOCS Official MethodCd 8-53. In *Official Methods and Recommended Practices of the American oil chemists' Society*.

Ayala-Garay, A. V., Espitia-Rangel, E., Rivas-Valencia, P., Martínez-Trejo, G., & Almaguer-Vargas, G. (2016). Análisis de la cadena del valor de amaranto en México. *Agricultura Sociedad y Desarrollo*. https://doi.org/10.22231/asyd.v13i1.280.

Aydeniz Güneser, B., Yilmaz, E., & Ok, S. (2017). Cold pressed versus refined winterized corn oils: Quality, composition and aroma. *Grasas y Aceites*. https://doi.org/10.3989/gya.1168162.

Ballesteros-Torres, J. M., García-Reyes, L. F., Fuentes-Tamez, A. G., Vara-Rangel, C. A., & Cabello-Treviño, L. A. (2016). Evaluación Microbiológica, De Rancidez Y Vida De Anaquel De Seis Envolturas Comerciales Utilizadas Para Empacado Al Vacío De Pierna De Cerdo [Microbiological, Rancidity and Shelf Life Evaluation of Six Commercial Wraps Used for Vacuum Packing of Pork Leg]. *Investigación y Desarrollo En Ciencia y Tecnología de Alimentos*, *1*(2), 158–166.

Banach, J. C., Clark, S., & Lamsal, B. P. (2016). Instrumental and Sensory Texture Attributes of High-Protein Nutrition Bars Formulated with Extruded Milk Protein Concentrate. *Journal of Food Science*, *81*(5), S1254–S1262. https://doi.org/10.1111/1750-3841.13270.

Barba de la Rosa, A. P., Fomsgaard, I. S., Laursen, B., Mortensen, A. G., Olvera-Martínez, L., Silva-Sánchez, C., ... De León-Rodríguez, A. (2009). Amaranth (Amaranthus hypochondriacus) as an alternative crop for sustainable food production: Phenolic acids and flavonoids with potential impact on its nutraceutical quality. *Journal of Cereal Science*, *49*(1), 117–121. https://doi.org/10.1016/j.jcs.2008.07.012.

Berke, T. G., & Shieh, S. C. (2012). Capsicum cultivars. In *Handbook of Herbs and Spices:* Second Edition. https://doi.org/10.1533/9780857095671.116.

Bressani, R. (2006). *Estudios Sobre la Industrialización del Grano de Amaranto, Caracterización Química y Nutricional de Productos Intermedios y Finales del Procesamiento*. Consejo Nacional de Ciencia y Tecnologia Guatemala [*Studies on the Industrialization of the Amaranth Grain, Chemical and Nutritional Characterization of Intermediate and Final Processing Products*. National Council of Science and Technology Guatemala]. Guatemala. Retrieved from http://glifos.concyt.gob.gt/digital/fodecyt/fodecyt 2002.23.pdf.

Cárdenas-Torres, F. I., Reyes-Moreno, C., Vergara-Jiménez, M. de J., Cuevas-Rodríguez, E. O., Milán-Carrillo, J., Gutiérrez-Dorado, R., ...

Cabrera-Chávez, F. (2019). Assessing the sensitizing and allergenic potential of the albumin and globulin fractions from amaranth (Amaranthus hypochondriacus) grains before and after an extrusion process. *Medicina (Lithuania).* https://doi.org/10.3390/medicina 55030072.

Cerecedo-Cruz, L., Azuara-Nieto, E., Hernández-Álvarez, A. J., González-González, C. R., & Melgar-Lalanne, G. (2018). Evaluation of the oxidative stability of chipotle chili (Capsicum annuum L.) oleoresins in avocado oil. *Grasas y Aceites*, *69*(1). https://doi.org/10.3989/gya. 0884171.

Cordle, C. T. (2004). Soy Protein Allergy: Incidence and Relative Severity. *The Journal of Nutrition*. https://doi.org/10.1093/jn/134.5.1213s.

Cornejo, L., Gaido, A., & López, C. (2017). *Snack a base de harina de amaranto con el agregado de spirulina, libre de gluten* [*Snack based on amaranth flour with the addition of spirulina, gluten-free*]. Universidad Nacional de Córdoba Argentina.

FAO. (2013). *2013 The International Year of Quinoa*. Retrieved April 29, 2020, from http://www.fao.org/quinoa-2013/press-room/news/ detail/en/.

Garza Benito, F., Ferreira Montero, I. J., & del Río Ligorit, A. (2006). Prevención y tratamiento del síndrome metabólico [Prevention and treatment of metabolic syndrome]. *Revista Española de Cardiologia*. https://doi.org/10.1157/13083448.

Gómez-Flores, Gómez-Acuña, G. A., Pérez-Díaz, V. H., & Chávez-Murillo. (2016). Desarrollo De Una Barra Nutritiva a Partir De Cereales Y Leguminosas: Análisis Proximal Y Sensorial [*Development of a Nutritional Bar from Cereals and Legumes: Proximal and Sensory Analysis*]. *Investigación y Desarrollo En Ciencia y Tecnología de Alimentos*.

He, X., & Rui, H. L. (2006). Cranberry phytochemicals: Isolation, structure elucidation, and their antiproliferative and antioxidant activities. *Journal of Agricultural and Food Chemistry*. https://doi.org/10.1021/ jf061058l.

Iwai, K., Suzuki, T., Fujiwake, H., & Iwai, K. (1979). Formation and Accumulation of Pungent Principle of Hot Pepper Fruits, Capsaicin and its Analogues, in *Capsicum annum* Var. Annuun Cv. Karayatsubusa at different Growth Stages after Flowering. *Agricultural and Biological Chemistry*. https://doi.org/10.1271/bbb1961.43.2493.

Jones, J. M. (2014). CODEX-aligned dietary fiber definitions help to bridge the "fiber gap." *Nutrition Journal*, *13*(1), 1–10. https://doi.org/10.1186/1475-2891-13-34.

Kalın, P., Gülçin, İl., & Gören, A. C. (2015). Antioxidant activity and polyphenol content of cranberries (Vaccinium macrocarpon). *Records of Natural Products*.

Lindsay, H. (1973). A colorimetric estimation of reducing sugars in potatoes with 3,5-dinitrosalicylic acid. *Potato Research*. https://doi.org/10.1007/BF02356048.

López-Legarda, X., Taramuel-Gallardo, A., Arboleda-Echavarría, C., Segura-Sánchez, F., & Restrepo-Betancur, L. F. (2017). Comparación de métodos que utilizan ácido sulfúrico para la determinación de azúcares totales. *Revista Cubana de Química*.

Melgar-Lalanne, G., Hernández-Álvarez, A. J., Jiménez-Fernández, M., & Azuara, E. (2017). Oleoresins from Capsicum spp.: Extraction Methods and Bioactivity. *Food and Bioprocess Technology*, *10*(1), 51–76. https://doi.org/10.1007/s11947-016-1793-z.

Neto, C. C. (2007). Cranberry and Its Phytochemicals: A Review of In Vitro Anticancer Studies. *The Journal of Nutrition*. https://doi.org/10.1093/jn/137.1.186s.

NMX-F-066-S-1978. (1978, November 3). Determinación de cenizas en alimentos [Determination of ashes in food]. *Diario Oficial de La Federacion*.

NMX-F-083-1986. (1986, July 14). Alimentos. Determinación de humedad en productos alimenticios. NMX-F-083-1986 Alimentos Determinación de Humedad en Productos Alimenticios [Foods. Determination of humidity in food products. NMX-F-083-1986 Food

Determination of Moisture in Food Products]. *Diario Oficial de La Federacion.*

NMX-F-615-NORMEX-2004. (2004, May 11). Alimentos - Determinación de extracto etéreo (Método Soxhlet) en alimentos. Método de prueba [Food - Determination of ethereal extract (Soxhlet Method) in food. Testing method]. *Diario Oficial de La Federacion.*

NOM-092-SSA1-1994. (1994, November 10). Bienes y Servicios. Método para la cuenta de bacterias aerobias en placa [Goods and services. Method for aerobic bacteria count in plate]. *Diario Oficial de La Federacion.*

NOM-111-SSA1-1994. (1994, May 10). Bienes y Servicios. Método para la cuenta de mohos y levaduras en alimentos [Goods and services. Method for counting molds and yeasts in food]. *Diario Oficial de La Federacion.*

NOM-113-SSA1-1994. (1994, May 10). Bienes y Servicios. Métodos para la cuenta de microorganismos coliformes totales en placa [Goods and services. Methods for counting total coliform microorganisms on plate]. *Diario Oficial de La Federacion.*

NOM-F-68-S-1980. (1980, August 4). Alimentos. Determinación de Proteínas [Foods. Protein Determination]. *Diario Oficial de La Federacion.*

Omolo, M. A. (2014). Antimicrobial Properties of Chili Peppers. *Journal of Infectious Diseases and Therapy.* https://doi.org/10.4172/2332-0877.1000145.

Orellana-Escobedo, L., Garcia-Amezquita, L. E., Olivas, G. I., Ornelas-Paz, J. J., & Sepulveda, D. R. (2013). Capsaicinoids content and proximate composition of Mexican chili peppers (Capsicum spp.) cultivated in the State of Chihuahua. *CYTA - Journal of Food.* https://doi.org/10.1080/19476337.2012.716082.

Rastogi, A., & Shukla, S. (2013). Amaranth: A New Millennium Crop of Nutraceutical Values. *Critical Reviews in Food Science and Nutrition, 53*(2), 109–125. https://doi.org/10.1080/10408398.2010.517876.

Restrepo-Gallego, M. (2007). Oleorresinas de capsicum en la industria alimentaria [Oleoresins of capsicum in food industry]. *Lasallista De Investigacion*.

Sanz-Penella, J. M., Wronkowska, M., Soral-Smietana, M., & Haros, M. (2013). Effect of whole amaranth flour on bread properties and nutritive value. *LWT - Food Science and Technology*, *50*(2), 679–685. https://doi.org/10.1016/j.lwt.2012.07.031.

Singh, N., Singh, P., Shevkani, K., & Virdi, A. S. (2019). Amaranth: Potential Source for Flour Enrichment. In *Flour and Breads and Their Fortification in Health and Disease Prevention*. https://doi.org/10.1016/b978-0-12-814639-2.00010-1.

Soni, R. A., Sudhakar, K., & Rana, R. S. (2017). Spirulina – From growth to nutritional product: A review. *Trends in Food Science and Technology*, *69*, 157–171. https://doi.org/10.1016/j.tifs.2017.09.010.

Soriano-Santos, J., & Avendaño-Martinez, A. (2013). El amaranto : alcances y limitaciones de sus propiedades tecnofuncionales para su diversificación [Amaranth: scope and limitations of its technical-functional properties for its diversification]. *Contactos*, *89*(May), 20–21.

Srichuwong, S., Curti, D., Austin, S., King, R., Lamothe, L., & Gloria-Hernandez, H. (2017). Physicochemical properties and starch digestibility of whole grain sorghums, millet, quinoa and amaranth flours, as affected by starch and non-starch constituents. *Food Chemistry*, *233*, 1–10. https://doi.org/10.1016/j.foodchem.2017.04.019.

Velásquez, R., & Llacho, K. (2015). *Aceptabilidad y valor nutricional de una barra nutritiva a base de harina de tarwi, kiwicha expandida y harina de trigo* [*Acceptability and nutritional value of a nutritional bar based on tarwi flour, expanded kiwicha and wheat flour*]. Universidad Nacional de San Agustin, Arequipa Perú.

Velastegui, A. (2016). *Desarrollo de un alimento nutritivo y energético tipo barra a partir de moringa, quinoa y amaranto* [*Development of a nutritious and energetic food type bar from moringa, quinoa and amaranth*]. Universidad de Guayaquil.

Vera Zambrano, M., Dutta, B., Mercer, D. G., MacLean, H. L., & Touchie, M. F. (2019). Assessment of moisture content measurement methods of dried food products in small-scale operations in developing countries: A review. *Trends in Food Science and Technology*. https://doi.org/10.1016/j.tifs.2019.04.006.

White, B. L., Howard, L. R., & Prior, R. L. (2010). Proximate and polyphenolic characterization of cranberry pomace. *Journal of Agricultural and Food Chemistry*. https://doi.org/10.1021/jf902829g.

WHO/FAO/UNU. (2007). Protein and amino acid requirements in human nutrition. *World Health Organization Technical Report Series*, (935), 1–265, back cover. Retrieved from http://www.ncbi.nlm.nih.gov/pubmed/18330140.

Yang, J., Pan, Z., Takeoka, G., MacKey, B., Bingol, G., Brandl, M. T., … Wang, H. (2013). Shelf-life of infrared dry-roasted almonds. *Food Chemistry*, *138*(1), 671–678. https://doi.org/10.1016/j.foodchem.2012.09.142.

INDEX

A

B

C

D

E

F

G

H

I

J

K

L

M

N

O

P

Q

R

S

T

U

V

W

Z